DISTRIBUTED COMPUTER CONTROL SYSTEMS 1983

Proceedings of the Fifth IFAC Workshop
Sabi-Sabi, Transvaal, South Africa, 18-20 May 1983

Edited by

M. G. RODD

University of the Witwatersrand
Johannesburg, South Africa

Published for the

INTERNATIONAL FEDERATION OF AUTOMATIC CONTROL

by

PERGAMON PRESS

OXFORD · NEW YORK · TORONTO · SYDNEY · PARIS · FRANKFURT

U.K.	Pergamon Press Ltd., Headington Hill Hall, Oxford OX3 0BW, England
U.S.A.	Pergamon Press Inc., Maxwell House, Fairview Park, Elmsford, New York 10523, U.S.A.
CANADA	Pergamon Press Canada Ltd., Suite 104, 150 Consumers Road, Willowdale, Ontario M2J 1P9, Canada
AUSTRALIA	Pergamon Press (Aust.)Pty. Ltd., P.O. Box 544, Potts Point, N.S.W. 2011, Australia
FRANCE	Pergamon Press SARL, 24 rue des Ecoles, 75240 Paris, Cedex 05, France
FEDERAL REPUBLIC OF GERMANY	Pergamon Press GmbH, Hammerweg 6, D-6242 Kronberg-Taunus, Federal Republic of Germany

First edition 1984

Library of Congress Cataloging in Publication Data

Main entry under title:
Distributed computer control systems 1983.
Proceedings of the IFAC Workshop on Distributed Computer Control Systems, sponsored by the International Federation of Automatic Control Computer Committee. Includes index.
1. Automatic control—Data processing—Congresses. 2. Electronic data processing—Distributed processing—Congresses.
I. Rodd, M. G. II. International Federation of Automatic Control. III. IFAC Workshop on Distributed Computer Control Systems (5th : 1983 : Sabi-Sabi, South Africa)
IV. International Federation of Automatic Control. Computer Committee.
TJ212.2.D57 1984 629.8'95 83-19494

British Library Cataloguing in Publication Data

IFAC Workshop *(5th : 1983 : Sabi-Sabi)*
Distributed computer control systems 1983 — (IFAC proceedings series)
1. Automatic control — Data processing — Congresses 2. Electronic data processing — Distributed processing — Congresses
I. Title II. Rodd, M. G. III. International Federation of Automatic Control IV. Series
629.8'95 TJ212

ISBN 0-08-030546-6

These proceedings were reproduced by means of the photo-offset process using the manuscripts supplied by the authors of the different papers. The manuscripts have been typed using different typewriters and typefaces. The lay-out, figures and tables of some papers did not agree completely with the standard requirements; consequently the reproduction does not display complete uniformity. To ensure rapid publication this discrepancy could not be changed; nor could the English be checked completely. Therefore, the readers are asked to excuse any deficiencies of this publication which may be due to the above mentioned reasons.

The Editor

Printed in Great Britain by A. Wheaton & Co. Ltd., Exeter

ANDERSONIAN LIBRARY
WITHDRAWN
FROM
LIBRARY
STOCK
UNIVERSITY OF STRATHCLYDE

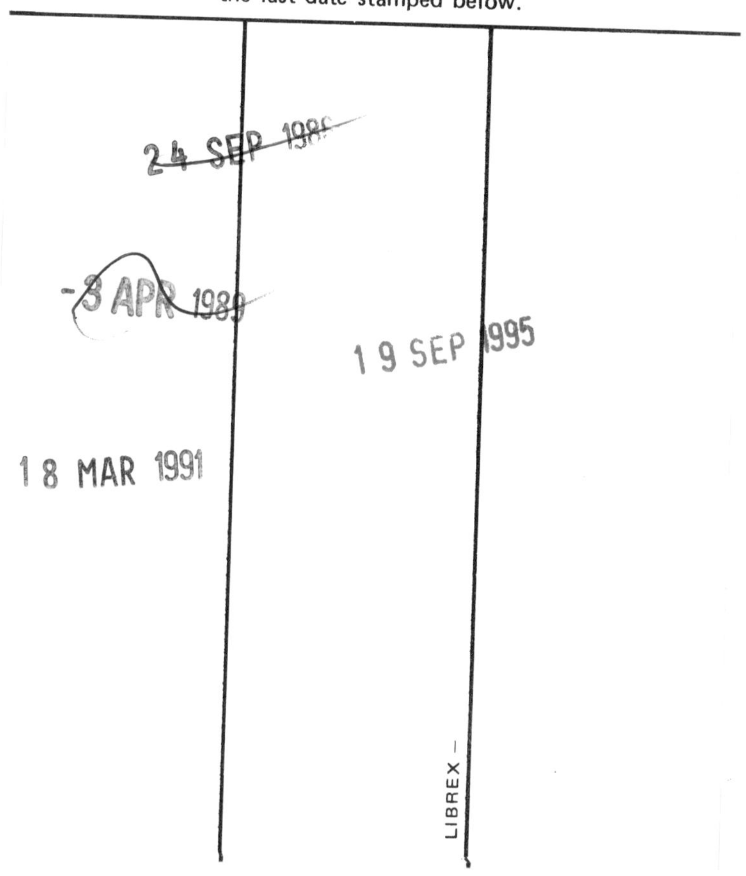
24 SEP 1989
-3 APR 1989
19 SEP 1995
18 MAR 1991
LIBREX

International Federation of Automatic Control

DISTRIBUTED COMPUTER CONTROL SYSTEMS 1983

Titles in the IFAC Proceedings Series

AKASHI: Control Science and Technology for the Progress of Society, 7 Volumes
ALONSO-CONCHEIRO: Real Time Digital Control Applications
ATHERTON: Multivariable Technological Systems
BABARY & LE LETTY: Control of Distributed Parameter Systems (1982)
BANKS & PRITCHARD: Control of Distributed Parameter Systems
BAYLIS: Safety of Computer Control Systems (1983)
BEKEY & SARIDIS: Identification and System Parameter Estimation (1982)
BINDER: Components and Instruments for Distributed Computer Control Systems
BULL: Real Time Programming (1983)
CAMPBELL: Control Aspects of Prosthetics and Orthotics
Van CAUWENBERGHE: Instrumentation and Automation in the Paper, Rubber, Plastics and Polymerisation Industries (1980)
CICHOCKI & STRASZAK: Systems Analysis Applications to Complex Programs
CRONHJORT: Real Time Programming (1978)
CUENOD: Computer Aided Design of Control Systems
De GIORGO & ROVEDA: Criteria for Selecting Appropriate Technologies under Different Cultural, Technical and Social Conditions
DUBUISSON: Information and Systems
ELLIS: Control Problems and Devices in Manufacturing Technology (1980)
FERRATE & PUENTE: Software for Computer Control
FLEISSNER: Systems Approach to Appropriate Technology Transfer
GELLIE & TAVAST: Distributed Computer Control Systems (1982)
GHONAIMY: Systems Approach for Development (1977)
HAASE: Real Time Programming (1980)
HAIMES & KINDLER: Water and Related Land Resource Systems
HALME: Modelling and Control of Biotechnical Processes
HARDT: Information Control Problems in Manufacturing Technology (1982)
HARRISON: Distributed Computer Control Systems
HASEGAWA: Real Time Programming (1981)
HASEGAWA & INOUE: Urban, Regional and National Planning — Environmental Aspects
HERBST: Automatic Control in Power Generation Distribution and Protection
ISERMANN: Identification and System Parameter Estimation (1979)
ISERMANN & KALTENECKER: Digital Computer Applications to Process Control
JANSSEN, PAU & STRASZAK: Dynamic Modelling and Control of National Economics (1980)
JOHANNSEN & RIJNSDORP: Analysis, Design, and Evaluation of Man-Machine Systems
LANDAU: Adaptive Systems in Control and Signal Processing
LAUBER: Safety of Computer Control Systems (1979)
LEININGER: Computer Aided Design of Multivariable Technological Systems
LEONHARD: Control in Power Electronics and Electrical Drives (1977)
LESKIEWICZ & ZAREMBA: Pneumatic and Hydraulic Components and Instruments in Automatic Control
MAHALANABIS: Theory and Application of Digital Control
MILLER: Distributed Computer Control Systems (1981)
MUNDAY: Automatic Control in Space
NAJIM & ABDEL-FATTAH: Systems Approach for Development (1980)
NIEMI: A Link Between Science and Applications of Automatic Control
NOVAK: Software for Computer Control
O'SHEA & POLIS: Automation in Mining, Mineral and Metal Processing (1980)
OSHIMA: Information Control Problems in Manufacturing Technology (1977)
PAU: Dynamic Modelling and Control of National Economies (1983)
RAUCH: Applications of Nonlinear Programming to Optimization and Control
RAUCH: Control Applications of Nonlinear Programming
REMBOLD: Information Control Problems in Manufacturing Technology (1979)
RIJNSDORP: Case Studies in Automation related to Humanization of Work
RIJNSDORP & PLOMP: Training for Tomorrow - Educational Aspects of Computerised Automation
RODD: Distributed Computer Control Systems (1983)
SANCHEZ & GUPTA: Fuzzy Information, Knowledge Representation and Decision Analysis
SAWARAGI & AKASHI: Environmental Systems Planning, Design and Control
SINGH & TITLI: Control and Management of Integrated Industrial Complexes
SMEDEMA: Real Time Programming (1977)
STRASZAK: Large Scale Systems: Theory and Applications (1983)
SUBRAMANYAM: Computer Applications in Large Scale Power Systems
TITLI & SINGH: Large Scale Systems: Theory and Applications (1980)
WESTERLUND: Automation in Mining, Mineral and Metal Processing (1983)
Van WOERKOM: Automatic Control in Space (1982)
ZWICKY: Control in Power Electronics and Electrical Drives (1983)

NOTICE TO READERS

Dear Reader

If your library is not already a standing/continuation order customer to this series, may we recommend that you place a standing/continuation order to receive immediately upon publication all new volumes. Should you find that these volumes no longer serve your needs, your order can be cancelled at any time without notice.

ROBERT MAXWELL
Publisher at Pergamon Press

IFAC Related Titles

BROADBENT & MASUBUCHI: Multilingual Glossary of Automatic Control Technology
EYKHOFF: Trends and Progress in System Identification
ISERMANN: System Identification Tutorials (*Automatica Special Issue*)

IFAC WORKSHOP ON DISTRIBUTED COMPUTER CONTROL SYSTEMS 1983

Organized by
The South African Council for Automation and Computation (SACAC)
The Council for Mineral Technology (MINTEK)

Sponsored by
The International Federation of Automatic Control Computer Committee (COMPCON)

PREFACE

One of the greatest values of holding a series of annual Workshops on a given topic is the way in which the growth of an idea can be highlighted at its various developmental stages. This has certainly been true of the IFAC Workshops on Distributed Computer Control Systems, and the Fifth in the series has upheld the tradition. Although attendance at each Workshop has been dominated by participants from the host country, there is a core of authors, of very varied backgrounds, who have presented papers which may appear superficially to be widely divergent. Nevertheless, particularly in the discussion sessions, a common thread of fundamental principles runs through.

The current Workshop was notable for the general agreement which became evident in a number of areas. Probably the most significant was the wide acceptance of the vital role of Real real-time in a distributed system. There are differing views on implementation, but it has become clear that Real real-time must be a fundamental consideration in the planning of all future DCCS architectures.

Communication systems, and the protocols involved, also received much attention. Again, there was, inevitably, some disagreement over issues such as broadcast messages and addressing techniques; also the types of messages required: state, event, immediate, consistent, etc. In spite of confusion over the exact requirements for a generalized approach, many common concepts are being developed. The actual communication media to be used provoked much discussion, and the work of the IEEE-802 Committee is having a serious impact - as is PROWAY.

Another aspect which continues to be of great interest is the language question, as increasing numbers of users are looking beyond FORTRAN. ADA is generally being hailed as a major development, despite some reservations over the interprocess communication primitives available.

Finally, the issue of fault-tolerance recurred, to the extent that many delegates to this event felt that a Distributed Computer Control System should, by definition, be fault-tolerant!

The Workshops have been noted for their high standard of papers, and the 5th Workshop maintained this level. The policy of inviting the bulk of the authors, and of accepting only a limited number of papers has paid dividends. Such presentations, in the Workshop situation, lead to fruitful discussion which is in many ways the highlight of the entire exercise. The Sabi Sabi venue, remote and highly unusual, provided an ideal environment for such debate. It is the Editor's hope that readers of the edited discussions contained in these Proceedings will benefit from that stimulating exchange of ideas.

CONTENTS

WELCOMING ADDRESS

G. Brown

President of the South African Council for Automation and Computation.
AECI Limited, PO Box 796, Germiston, South Africa

It is my sincere pleasure, to welcome you all to the 5th IFAC Workshop on Distributed Computer Control Systems. I especially welcome our overseas delegates who have taken the time and trouble to travel great distances to join us here at Sabi-Sabi.

It seems to me that it is most appropriate that this Workshop should be held in South Africa at this time. Whilst South Africa is clearly a developing country and is still small in the industrial sense, we are, I believe, a technologically progressive nation. For example, I think we can lay claim to one of the largest installed proprietary Distributed Control Systems anywhere in the world, referring, of course, to the Sasol II and III oil from coal plants. For those of you with knowledge of these huge plants, it is inconceivable to think that they could be successfully controlled without the facilities provided by Distributed Computer Control Systems. Several other processing industries in this country are also users of Distributed Control Systems on a significant scale.

In parallel with this growing local industrial application of proprietary DCCS, the past few years have seen the vigorous growth of important research work in this field at our Universities and other institutions and notable results have already been achieved by this research. Thus, we now have in South Africa the beginning of an excellent environment for cross fertilisation of ideas between experienced and imaginative industrial users and highly competent research and development teams.

The Sabi-Sabi Workshop thus provides a timely platform for many South African workers in this field to exchange ideas with our esteemed overseas delegates and thus add an international dimension to the beginning of the local environment of cross fertilisation of ideas.

A Workshop of this type does not occur without considerable effort being expended by many people, and I cannot let this opportunity pass without expressing my thanks to all concerned. In particular I must thank the authors for their work in preparing the papers, without which the Workshop could not take place. Our thanks are also due to the International Program Committee, under the Chairmanship of Warren Gellie for selecting a set of papers which are clearly of high standard. Finally, on behalf of SACAC, I would like to thank Mike Rodd and his National Organizing Committee for a tremendous job well done!

THE DISTRIBUTED DATA FLOW ASPECT OF INDUSTRIAL COMPUTER SYSTEMS

R. Güth and Th. Lalive d'Epinay

Brown Boveri Research Center, CH-5405 Baden, Switzerland

Abstract. Today's computer systems are in general based on the von Neumann type architecture. However, there also exist alternative architectures, for example dataflow systems and (quasi-) continuous systems.

This paper presents a new type of computer architecture which is based on broadcast principle and on source addressing. This system can be viewed as a combination and generalisation of dataflow and continuous systems. It has some inherent advantages over the classical architecture, but there are also some difficulties in efficient implementation. A concept has been established which allows to implement broadcast systems efficiently and hence to utilize their advantages.

There already exist distributed process control systems based on the broadcast and source addressed architecture. Ongoing research work in the field of fundamental concepts and their efficient implementation will broaden the scope of these systems and make them applicable in a wide range of process control and other applications.

Keywords. Dataflow system; control system; continuous system; broadcast system; source addressing; communication ether.

A DATAFLOW VIEW OF CONTROL SYSTEMS

Most of today's computer systems for commercial as well as for process control applications are based on a conventional, von Neumann type architecture. As will be shown in this paper, an alternative architecture has many advantages. We will present a logical model of an architecture that is derived from and will cover two extremes:

1) Dataflow systems: This concept is mainly known and propagated by a relatively small group of computer scientists. Despite its theoretical advantages it has only been practically used for special applications.

2) Continuous systems (analogue computer): This concept is well known by control engineers. The introduction of digital processors has pushed away this concept at least from the realization level of computer control systems.

Dataflow System

A dataflow language has no concepts of variables and explicit control flow: a program is a set of data activated functions connected by unidirectional data paths. In a dataflow program the order of function executions is not defined explicitely (by the control structure of a procedural program), but derived implicitely from the dataflow that is related to the computation (Ackermann, 1982; Agerwala, 1982; Davis, 1982; Dennis, 1979; Treleaven, 1982).

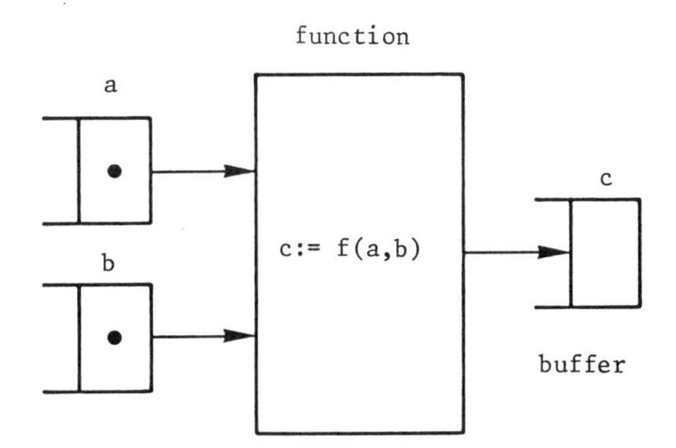

Fig. 1: Element of a dataflow program
The dots (a and b present, c absent) represent the condition which allow firing of the operation f

Fig. 1 shows an element of a dataflow program, where an output c is computed by execution of a function f(a,b). The execution of a function consumes all input values and produces an output value. A function is executed only if all input values exist (have been produced) and the output is empty (has been consumed). This execution condition is called firing rule. A dataflow program is defined by interconnecting several functions.

Function interconnections can be considered as buffers. By writing a result value a buffer content is produced and by reading a buffer its content is consumed. Function executions are controlled by the associated buffers, i.e. function execution is data driven.

Continuous System

In a continuous system an output funtion is continuously computed from an input function. As long as the boundary frequency of the operator is sufficiently high compared to the relevant frequencies of the signals, a continuous system is an ideal tool for control and simulation purposes. It is a natural extension of a physical process.

A continuous system can be regarded as a dataflow system with continuous firing or in a first approximation, with periodic firing with a sufficiently small period (Fig. 2).

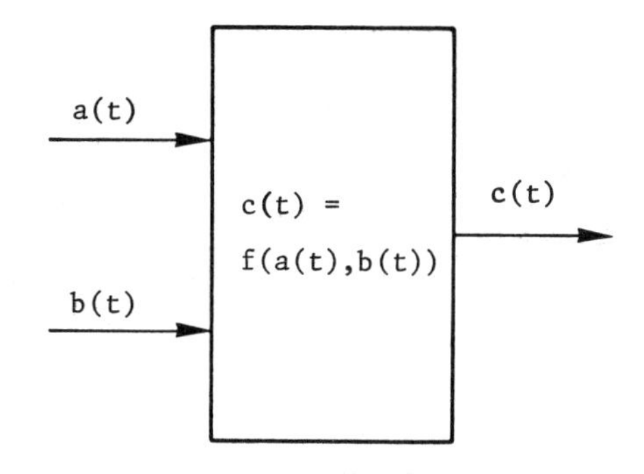

Fig. 2: Continuous System

Quasi-Continuous Systems

Some modern microcomputer-based distributed control systems can be considered as quasi-continuous systems. In such systems continuous control functions are realized approximately by algorithms and executed on microcomputer-based controllers in sufficiently short cycles. There is usually no need for a strict producer/consumer relationship of values, as it is found in dataflow systems. Rather, function executions are normally time cycle driven and function results update previously generated values. That is, functions communicate via (memory) cells: by writing a new value into a cell the cell content is updated, by reading a cell the cell content is not destroyed (Fig. 3).

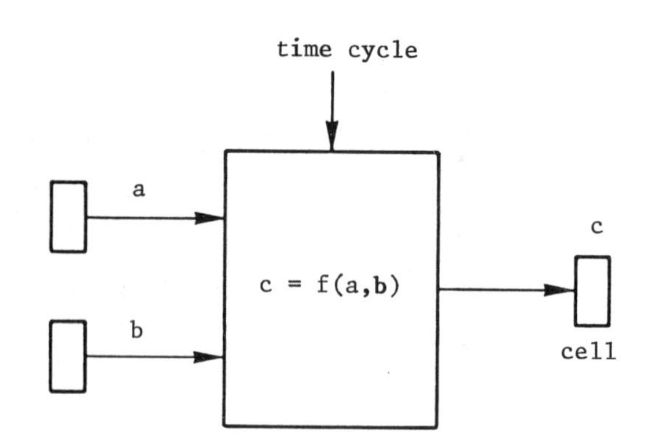

Fig. 3: Quasi-continuous system

Examples

For the sake of illustrating the application of a (pure) dataflow machine and a quasi-continuous system let's have a short look at two typical applications:

1) bank transaction
A bank transaction can be implemented by a dataflow program with the implicit firing rule: the input values must be present, they are consumed and produce an output value. The operation has to be executed exactly once, it may not be repeated nor left out (Fig. 4).

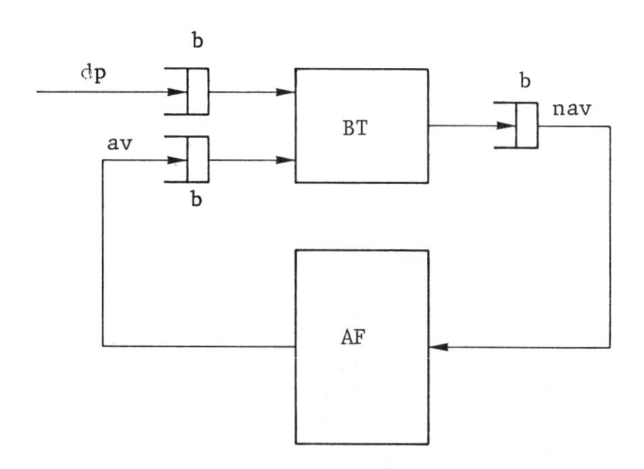

Fig. 4: Symbolic representation of a bank transaction as a dataflow program.

BT = bank transaction
AF = access functions of the account data base
b = buffer
dp = deposit
av = account value
nav = new account value

2) process control
A PID control algorithm is a typical example of an operation, which can be executed quasi-continuously (or as often as possible, using an appropriate integration and differentiation method). (Fig. 5)

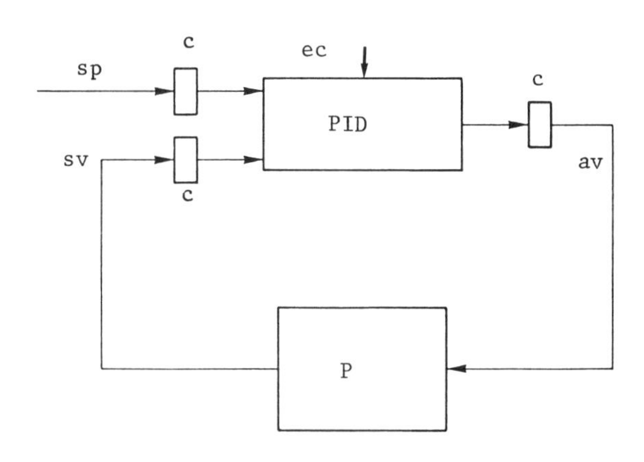

Fig. 5: Representation of a control function in a quasi-continuous system

PID = PID control function
P = physical process
c = cell
sp = set point
ec = execution cycle
sv = sensor value
av = actuators value

BROADCAST SYSTEMS

Motivation

In the following we introduce a type of distributed computer system that has some features in common with dataflow systems. We call these systems Broadcast Systems.

The motivation for broadcast systems is to provide an operational principle that facilitates the construction/configuration of large software systems from predefined building blocks (programming in the large). The building blocks are assumed to be represented by program modules that are internally implemented in procedural high-level languages (programming in the small).

The building blocks from which large programs are constructed, are expected to be maintained in a module catalogue. The user is free to enhance the catalogue by user-defined module types. Modules can represent simple functions, type managers (abstract data types), etc. (see Fig. 6). The catalogue provides module types from which instances are created and added to the already existing part of the software system (see Fig. 7). At program runtime the modules are equipped with work space and can be considered as processes, monitor processes, type managers, etc.

We should recall here the preconditions for an extensive use of predefined modules:

- The modules have to be independent from the environment, i.e. in particular from the hardware and software configuration.

function

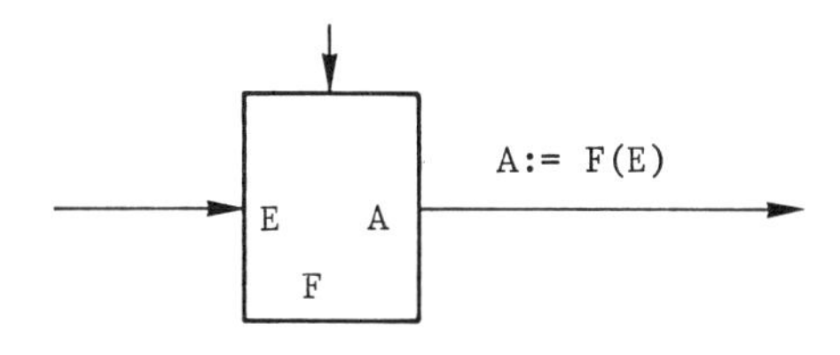

process

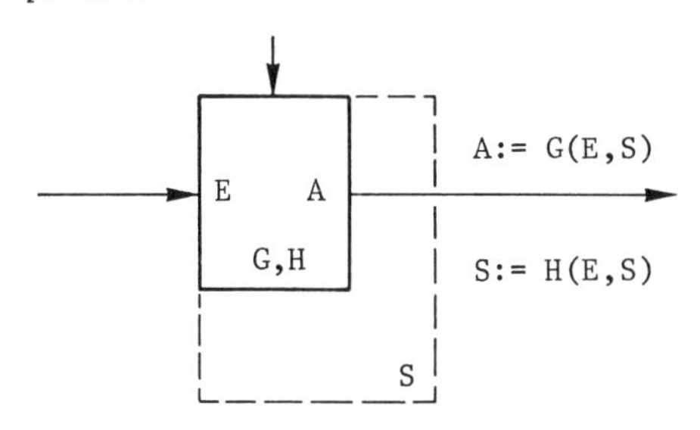

type manager (abstract data type)

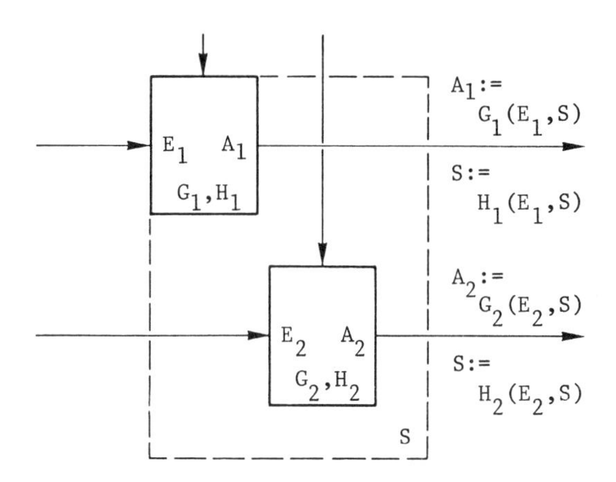

Fig. 6: Examples of program modules

E = input
A = output
S = internal state
G, H = functions

- A uniform framework has to be provided for the interconnection of modules.

The second precondition is a strong demand for appropriate computer architectures.

Basic Ideas

The basic concept of broadcast systems is a communication ether which provides all resources necessary for the transmission of values. All values have a unique source and are available wherever needed.

For the sake of simplicity we will restrict the following discussion to simple functions

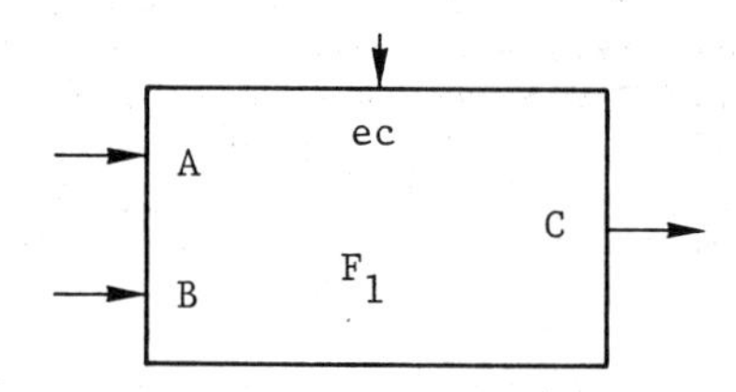

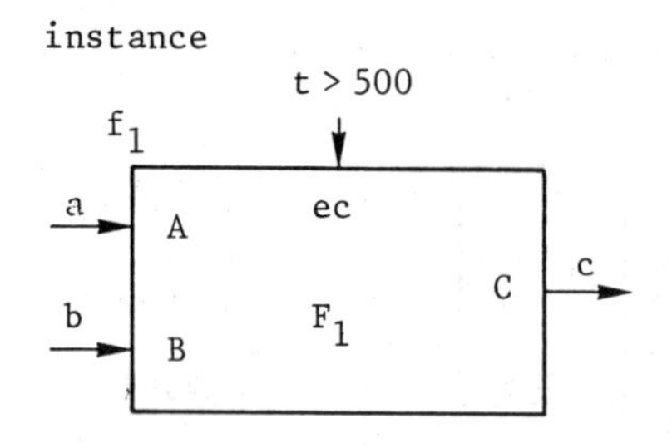

Fig. 7: Creation of an instance from a function type

F_1 = function type
A,B,C = formal parameters
f_1 = function instance
a,b,c = actual parameters
ec = execution condition

instead of general modules. To each function an execution condition is associated that defines the precondition for the execution of the function. The execution condition can be expressed in the form of a boolean expression on all data available in the ether including time. If and only if the execution condition is satisfied the associated function will be executed (Fig. 8).

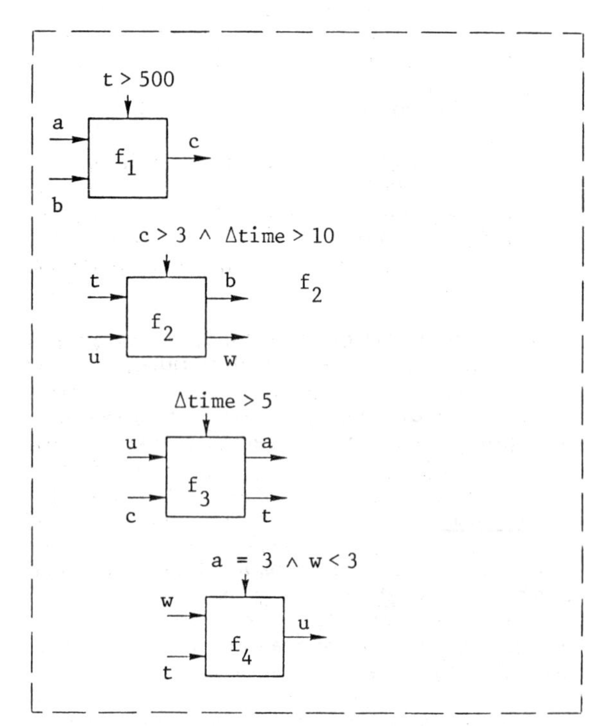

Fig. 8: Communication ether

Data, that are results of function executions, are broadcast in messages with source identification. Thus, data are globally accessible within the ether. All functions listen to transmitted messages and receive only those that are required for their inputs.

Let us summarize the advantages of that simple and robust operational principle for the interconnection and cooperation of functions and program modules:

- there are no side effects of function executions
  - a side effect is the change of a value that is not explicitely associated with an output parameter of a function
- there is no risk in having all values globally known
  - there is no need for scope rules and for passing access rights
- the approach facilitates the
  - incremental construction of software systems,
  - extension of running software systems
- there is a natural support of backward error tracing and the identification of fault sources by
  - source identification of values,
  - definition of execution conditions

Fig. 9 illustrates the extension of an existing program, given in Figure 8, by an additional function. Obviously, the introduction of the new function does not affect the original program.

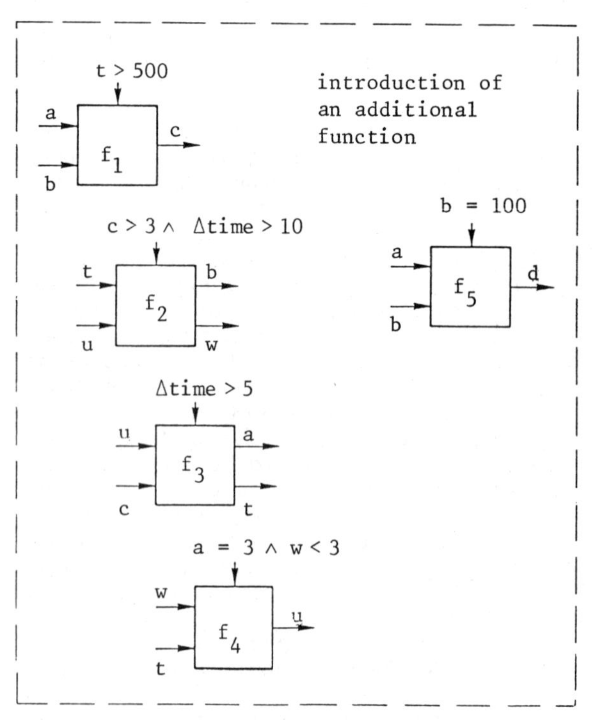

Fig. 9: Extension of the program given in Fig. 8

Let us introduce another model that is partly equivalent to the communication ether: the source addressed memory. The source addressed memory is similar to a conventional memory, with the exception that each cell is associated exclusively to one source, i.e. to one writer. A cell can be read by arbitrary functions of the program, using the appropriate source identification.

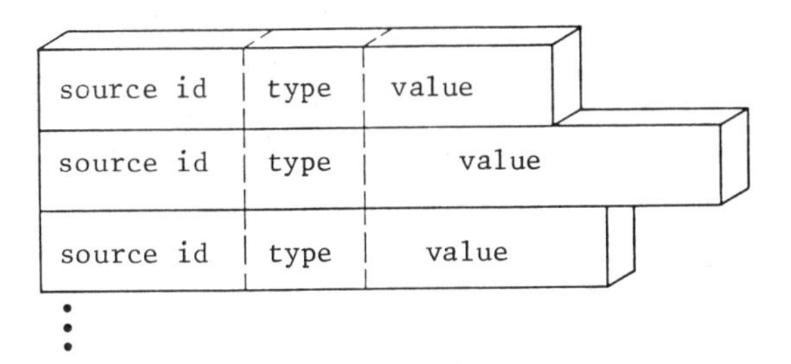

Fig. 10: Model of source-addressed memory

We present the model of source-addressed memory to emphasize the differences between function cooperation in the small and in the large. Conventional memories provide highly flexible and unrestricted means for storing and exchanging information. Since addresses can be created in unrestricted ways, cells of conventional memories can be changed from everywhere in the program system. Thus, side effects cannot be prevented and fault sources are difficult to localize. Features that should be tolerated only within small program units.

We feel that different mechanisms should be provided for programming in the small and software function interconnection in the large.

Realization of Broadcast Systems

A communication ether can be realized straightforwardly in distributed computer systems. Results of function execution are broadcast in messages with source identification. These messages are sent over a communication network and, thereby, are made available to all node computers connected to the network.

Within the node computers dedicated functional units perform particular operations on the communication ether:

- An associative receiver listens to all the messages on the network and copies only those data in its internal memory that are needed by the node computer

- An execution condition evaluator permanently evaluates the execution conditions of the functions installed on the node computer.

- A transmitter broadcasts the results of function executions. The broadcast mechanism supports consistent, simultaneous updates of memory cells at different places. The commitment of result values is atomic, i.e. either all receiving node computers perform an update or no node does.

The interconnected functions are executed on processors that are part of the node computers. Fig. 11 gives an impression of the internal structure and organization of the node computers.

We should point out that message broadcast allows effective use of the communication network because one 1:n transmission replaces n 1:1 (point to point) transmissions. The execution time overhead of associative message receiving and execution condition evaluation is kept small by the dedicated hardware units.

The key issue of the communication system is: how to establish acknowledge of broadcast messages. Since the number of receivers is both variable and unknown, the solution is the logical or of negative acknowledgements. That raises a second question: how to detect a dead receiver that cannot even send a negative acknowledge. That problem is solved by letting each node computer send its negative acknowledge signal regularly. This is then checked by one or more dedicated supervising units.

Programming Environment

As the control-engineer is used to formulate his problem in a way very similar to a dataflow program, he programs the system graphically, using off-site CAD-systems or on-site interactive programming stations, which become more and more efficient and economical.

It is also important to realize that it is possible to convert automatically and unambiguously a machine program into a graphic representation and vice-versa without the need of any additional, auxiliary graphical information.

- The module types are realized internally in procedural high level languages and compiled into machine code over micro code for high-performance functional units. This is the working field of the system specialist. It is "programming in the small", where the overall size of the program is limited, independent of all others and of course also application-independent. This allows to optimize implementations of the operations.

- The application system, which may be distributed over more than 10'000 functional units, is programmed by the application engineer in a non-procedural, possibly graphical language as described

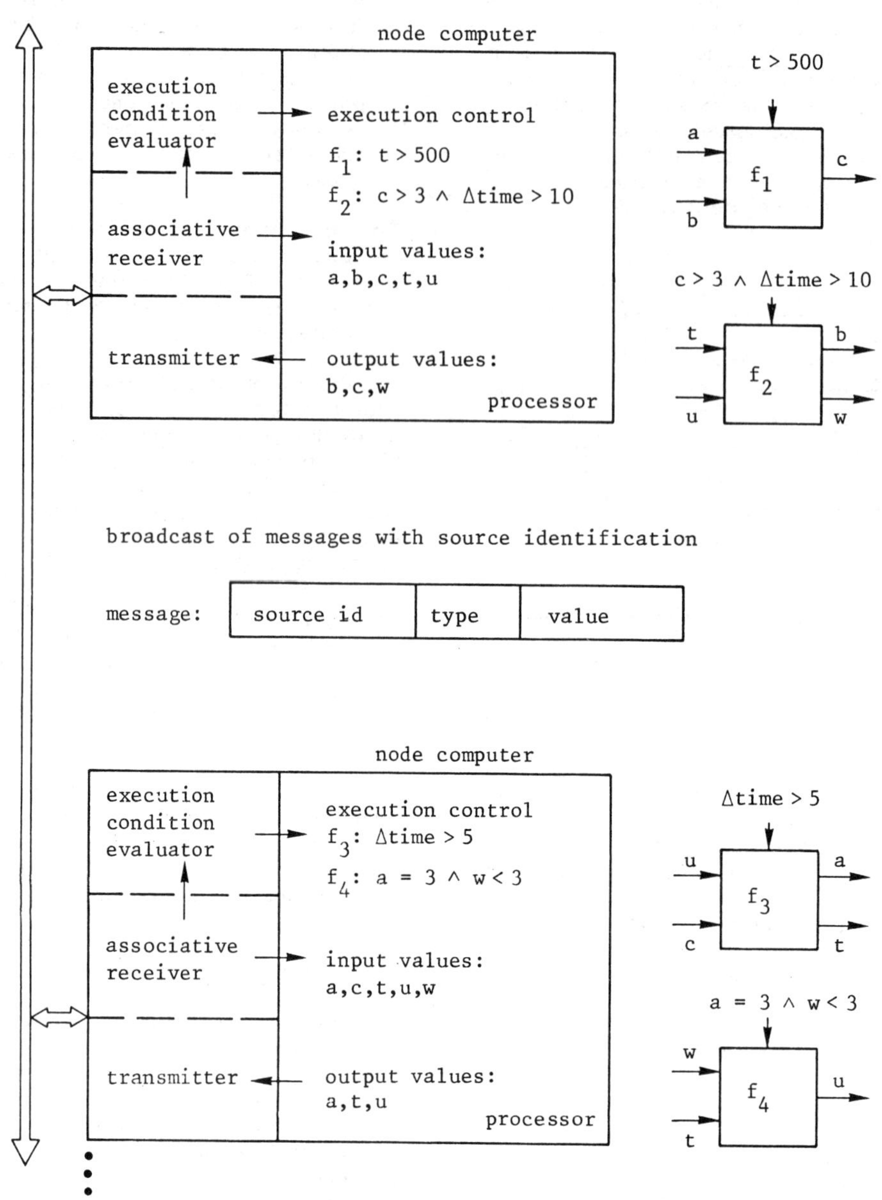

Fig. 11: Realization of broadcast systems

above ("programming in the large"). The principle of source-identification makes it possible that the distribution of the program over the physical modules (functional units) is totally transparent and can change, even while the system is running. This flexibility is very helpful, because it allows to write the application software before the hardware configuration is known. The advantage for reconfiguration and expansion of control system is obvious.

PERSPECTIVES

We have presented the concept of broadcast systems. Even in process control systems the full potential of the concept has not yet been totally used. Some additional and very precise work has to be done to prove the usefulness of the concept for a much wider range of applications. The field of process control, hopefully in a much wider range than up to now, will remain the main application of the concept at least for a certain time.

Corresponding research work is done and the results are expected to provide major impact for future systems.

PRESENT SYSTEMS

A consequent implementation of broadcast aspects has been realized in different products of Brown Boveri & Co., e.g. in the systems PROCONTROL, PARTNERBUS and BBC-Kent P4000. This is not the place to

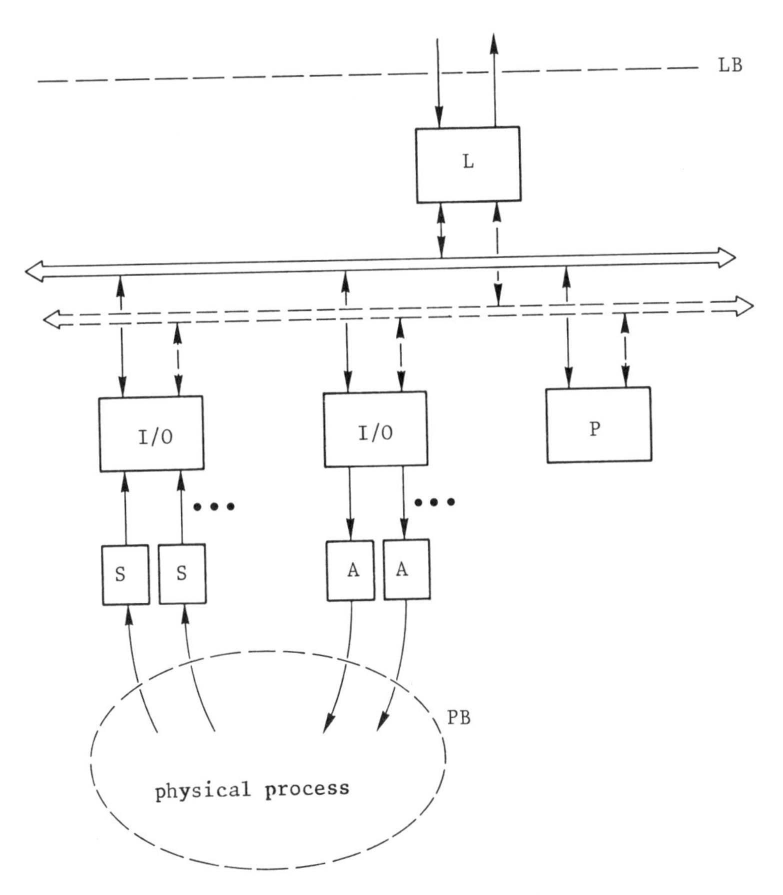

Fig. 12: Simplified view of the BBC-PROCONTROL System

S = Sensor
A = Actuator
I/O = Process Input and/or Output, including control and/or data processing functions
P = Processing Unit (Process Control, DDC, local man-process interface, local man-computer interface)
PB = physical boundary
LB = logical boundary
between process and control system

give a detailled product description; there exists an appropriate documentation. As these products are mainly used by the companies' own application departments and because the advantages of the broadcast concepts are not expected to be widely known in the field of process control applications these systems are not explicitely known as broadcast systems. Nevertheless, their inherent advantage over seemingly similar systems is based on this concept.

The systems are specialized according to the hierarchical structure of a process control system.

The system PROCONTROL is optimized to handle process data and to communicate them to and from programmable controllers (representing node computers) and process data input/output equipment.

The communication ether realized by the system PROCONTROL has the effect of moving the logical boundary between physical process and computer control system into the computer control system (Fig. 12). This means that sensor values, independent of the location of their generation, are available wherever needed in the control system and that the actuator values, again independent of the module of their computation, will reach the appropriate actuators. All functions of the system are free of side effects; the system can be programmed without knowing how the functions will be distributed among the physical modules and where these modules will be placed. The system itself is structured to allow efficient solutions for systems ranging from relatively small to very large applications. (The largest system being installed contains more than 5'000 node computers.)

All modules of the system contain associative receivers and transmitters. The communication within the system as well as the execution of the functions is oriented towards efficient quasi-continuous behaviour: The signals are communicated whenever they have changed; the amount of change which triggers a broadcast can be selected and is combined with minimal and maximal delays, so that a signal on one hand does not saturate the communication network and on the other hand is sent regularly, even if there was no or not sufficient change.

The communication protocol includes a common acknowledge of all connected modules as previously described.

The system includes some interesting features: it can be built up completely redundant without need of any specific programming; if the system is cut the separated parts immediately begin to work independently.

The PARTNERBUS system is oriented towards the communication of minicomputer-like and mainframe-like node computers. It implements the communication functions of the presented broadcast system, i.e. associative receiver and transmitter. The node computers can set up tables which define the signals that have to be received and the location in memory where this information has to be stored. They also define the names and values of the outgoing signals. The broadcast aspect can be restricted to the communication system to allow optimal modularity between nodes that internally are used conventionally. This allows even integration of a Partnerbus system into existing systems without changing their original operating systems (e.g. VAX/VMS*).

The BBC Kent P4000 System makes use of the broadcast concept to achieve a extremely flexible combination of packaged subsystems. This allows to tailor systems to the customer's need with minimal cost.

CONCLUSION

The concept of a broadcast architecture has some potential advantages over conventional systems. The practical exploitation of the concept has so far been believed to be restricted to a few special applications. In the field of process control the concept has been used successfully and corresponding products exist; a more general use is not unlikely in the the future.

ACKNOWLEDGEMENT

We want to express our gratitude to our colleagues in the research center, whose contributions were very helpful and constructive. In particular J. Kriz and S. Züger have directly contributed to this work and are involved in ongoing activities.

---

*) VAX and VMS are trademarks of DIGITAL EQUIPMENT CORPORATION

REFERENCES

Ackermann, W.B. (1982). Data Flow Languages. Computer, Vol. 15, No. 2, pp. 15-25.

Agerwala, T., Arvind. (1982). Data Flow Systems. Computer, Vol. 15, No. 2, pp. 10-13.

Davis, A.L., Keller, R.M. (1982). Data Flow Program Graphs. Computer, Vol. 15, No. 2, pp. 26-41.

Dennis, J.B. (1979). The Varieties of Data Flow Computers. Proc. 1st Int. Conf. Distributed Computing Systems, New York, IEEE, pp. 430-439.

Treleaven, P.C., Brownbrigde, D.R., Hopkins, R.P. (1982). Data-Driven and Demand-Driven Computer Architecture. ACM Computing Surveys, Vol. 14, No. 1, pp. 93-143.

DISCUSSION

Sloman: I can see a possible problem in using global names. Firstly, you appear to lose some abstraction, and secondly, you have a problem of managing your name-space. If you have to manage global source identifiers in a very large system, such as you refer to in your paper, it appears to be very difficult to choose names which are unique. How can this be overcome?

LaLive d'Epinay: This does not present any problem. The Power Systems people, for example, have a naming concept which is older than computer control systems, in which the name of a variable is uniquely defined by the geographical location in the building as well as the precise location of the place where the signal is created. In such a system there is no problem, since, from the beginning of the design, it is totally clear what the name of any variable is. They have a very large name-space in which they use only a very few possibilities: this is one of the reasons why the systems which we have developed have 24-bits of address space on the communication bus system. There are, however, possibly other areas of application where there is a problem. One of these arises in the case that I did not mention in my presentation, and that is the recognition of duplicate source identifications. This is typically built into the hardware and you can try out the name if you want, and the system will be able to tell you whether the name is already used or not. Of course the whole management of variable names and the names of signals on the cables has to be planned as part of the integral solution.

MacLeod: You have told us that in your approach there is a unique source for each piece of information. Does this not eliminate one of the advantages of distributed control systems, and that is redundancy? In other words, the ability to handle redundant senders of information in the event of a failure?

LaLive d'Epinay: This is also one of the problems which is solved by our solution. Let us say that you have redundant information. In this case there must be some knowledge of this fact. Let us say, for example, that a twin of the information exists. Recognition of this redundancy is built into the hardware - you can even have, for example, the typed-description of one of the products and you can see whether it is the original, whether it is a back-up or whether it is an engineer-supplied auxiliary value. For example, if the measurement of a temperature is not functioning you could take the redundant information as the temperature. You can recognize that in the data field, but, of course, you must know that twins or triplicates exist. Another aspect is that in a truly redundant system, the sensors are duplicated so you effectively have two different values right from the very beginning. They are treated differently. We can also have a redundant communication system which communicates every piece of information in a duplicate form so that we have the same information and the same source identification on both of the communication systems. It is undoubtedly a very complex problem which has to be solved on each level appropriately, but in our solution it can be approached in a very elegant fashion.

Sabi-Sabi, South Africa, 1983

REAL TIME IN DISTRIBUTED REAL TIME SYSTEMS

H. Kopetz

Institut für Praktische Informatik, Technische Universität, Wien, Austria

Abstract . Any real time computer control system must have a capability to measure the duration between events in the metric of real time and must respond to a stimulus within a given real time interval. This paper discusses some of the implications which result from the inclusion of this real time metric on the specification, communication and error detection in real time distributed systems.

Keywords . Real Time, Logical Time, Distributed Real Time Systems, Communication, Error Detection.

INTRODUCTION

The tremendous growth of decentralized computers has brought the topic of coordination and synchronization of the fairly autonomous systems to the center of interest. But, although synchronization is closely related to the activity of the systems in the domain of real time, many published synchronization techniques try to abstract from the property of real time at the earliest possible instance and are only concerned with the logical ordering of events /1,2,3,4/. In distributed real time systems such an abstraction is neglecting an essential property of the problem. Since real time can be a critical resource and the maximum duration of a response to an external event is dictated by the environment, the explicit consideration of the metric of real time is absolutely necessary in real time control systems.

This short paper, which has to be seen as a contribution to a workshop on Distributed Computer Control Systems. tries to discuss the following topics in DCCS from the point of view of real time

- the establishment of a time reference
- specification
- communications protocols
- error detection and
- state restoration

THE ESTABLISHMENT OF A TIME REFERENCE

In the following discussion we distinguish between logical time and physical time.

Def.: Logical Time
A time reference established by counting specified significant events (logical ticks). There exists no metric to

measure the interval between logical ticks. Thus the concept of duration between logical ticks is not defined.

Def.: Physical Time
A time reference established by counting the ticks of a physical clock (physical ticks). There exists in principle a metric to measure the interval between any two physical ticks (physical real time). The duration of these interval is, expressed in this metric, constant.

We also distinguish between a relative and absolute time reference:

Def.: Relative Time Reference
The time reference (logical or physical) is established in relation to a local event.

Def.: Absolute Time Reference
The time reference (logical or physical) is established in relation to a global event for a given system. The time reference for the absolute physical time is the UTC (Universal Time Coordinated).

We thus have the following options for the establishment of a time reference:

	start event	
time base	global	local
physical	UTC (global real time)	local real time
logical	global logical time	local logical time

In order to solve resource sharing problems in a distributed system it is necessary to establish a global time reference for all units concerned. Many systems create this global time reference by the synchronization of local logical clocks. Pratically every operational system (real time or not) operates with local physical timeouts. Therefore many distributed systems contain a global logical time as well as a number of local physical time bases. One can imagine that the coordination of these singular time references can give rise to severe conflicts.
Since in real time systems the maximum duration between physical events of the environment is given in the metric of real time, some physical time reference has to be available. In a DCCS it is in our opinion reasonable to establish not only a relative physical time reference but to introduce an absolute physical time reference which can be used for the control of access to global resources as well. The implementation of this global physical time reference in a distributed real time system is not trivial. The normal approach is the provision of local real time clocks at each node and the periodic resynchronization of these clocks to some global time base. Since there is always a finite delay between a signal sent from one node to another node and every clock has its individual physical pace, multiple clocks will always deviate by a small amount, i.e. there is always a point in realtime at which one clock has ticked already but another clock has not ticked yet. There are two solutions to this problem:

(1) to cease all system activity during this finite interval of uncertainty
(2) to develop protocols which are tolerant to clock deviations by one tick. One tick is sufficient since it is always possible to stretch the duration of a tick in order to cover the deviation between any two clocks in a system.

In our MARS /5/ implementation we used the second strategy.

SPECIFICATION

As mentioned before, real time systems are concerned with the provision of timely responses to requests from the environment. In addition to the functional specification, the specification of these necessary response times must be part of the problem statement.
It is our opinion that only behavioural specification techniques provide the expressive power needed to specify these response time requirements.
In real time control, the validity of information is cancelled by the passage of time, e.g. a measured variable which has not been updated for some time becomes suspect. There are many situations where the realization that no valid information is available is less of a problem than the unsuspecting use of outdated information. Any information item is thus valid only for a given time interval , which must be part of the specification of the item. After this time the information item becomes part of the history and shall not be used for control purposes any more. In distributed real time systems this problem is aggravated by the existence of an unreliable communication medium. If the loss of a message cannot be detected by the receiver - which is normally the case if PAR protocols are applied - then it is difficult to detect outdated information the validity of which is not limited in the time domain. To my knowledge there are no specification techniques which adequately address this problem of time validity of information.

COMMUNICATION

In many distributed systems the abstraction of a reliable connection between two partners is realized by application of time redundancy. Since in real time situations time can be a critical resource the generous use of this resource can give rise to a number of problems (e.g. in the PAR - Positive acknowledgement or retransmission - protocol):

- The time has invalidated the information before the communication protocol is successful. In this situation the communication system tries to transport outdated information.

- The time redundancy is only activated after an error has been detected (e.g. by timeout). The communication traffic is thus dependent on the communication failure rate and becomes unpredictable. Since the transportation delay is a function of the traffic, the transportation delay becomes also unpredictable.

- The error detection in the PAR protocol is realized at the sender, not at the receiver of information. In many process control situations error detection at the receiver, not at the sender, is required.

It is therefore our opinion that protocols of the PAR type - these are the most common protocols used in distributed systems - are not well suited for distributed computer control systems. A new class of protocols, which have the following characteristics is required in real time systems:

- predictable performance, even under error conditions
- error detection at the receiver of information
- n to n communication topology

Under real time conditions the abstraction of a "reliable connection", which is provided by the communication system might not be affordable.

ERROR DETECTION

In real time control systems error detection can be realized in two dimensions: the dimension of values and the dimension of time. Classical error detection techniques are mainly concerned with the dimension of values, e.g. the result of a computation must be consistent with a given output assertion. As already mentioned before, in addition to the value attribute every information item is also characterized by a time attribute. The systematic monitoring of this time attribute can be the most effective means of error detection in real time systems. It has been reported from practical experience /6/ that up to 99 % of all error conditions can be detected by comparing the actual and expected timing patterns of a computation. The effective use of this error detection mechanism requires the establishment of a common time reference and the attachment of the time attribute to every information item.
An appropriate real time operating system will support this time management and time monitoring, so that the task of the application programmer is reduced to specifying the application semantics and the application timing.

STATE RESTORATION

State restoration is concerned with the creation of an internal state of a computing system which is consistent with the external state of the environment. Two different techniques for state restoration are discussed in the literature

(1) backward recovery
(2) forward recovery

Backward recovery refers to the restoration of an internal state which was consistent some time ago (backward) and the subsequent modification of this state (or the environment) in order to establish the consistency at the present point in time.

Forward recovery refers to the creation of a new internal state which will be consistent with the environment sometimes in the near future (forward).

If there is no close interaction between a computer system and its environment state restoration by backward recovery can be implemented without problems.

We call the point real time in which the control over the results of a computation is passed to the environment the point of commitment. Every point of commitment invalidates all checkpoints which have been taken before the point of commitment. Only the checkpoints taken in the interval < point of commitment, present > remain intact.

Since in real time systems there are many points of commitments state restoration by backward recovery is severely limited. In our opinion state restoration by forward recovery is a better suited technique for DCCS.

CONCLUSION

Many of the positions taken in this note are rather provocative and unconventional. We hope that these views will be challenged in a lively discussion in order to reach a better understanding of the role of real time in distributed real time systems.

LITERATURE:

/1/ Lamport, L., Time, clocks and the ordering of events in a distributed system, CACM, Vol. 21, No. 7, July 1978

/2/ Le Lann, G., An Analysis of ifferent Approaches to Distributed Computing , Proc. 1st International Conference on Distributed Computing, Huntsville, October 1979

/3/ Schwartz, R.L., Melliar Smith, P.M., From State Machines to Temporal Logic, Specification Methods for Protocol Standards, IEEE Trans. on Comm. Vol-Com-30, Vol 12, Dec. 1982

/4/ Girault, C., Reisig, W., Application and Theory of Petri Nets, Informatik Fachberichte, Springer Verlag, Heidelberg 1982

/5/ Kopetz, H., Lohnert, H., Merker, W., Pauthner, G., The Architekture of MARS, Techn. Univ. Berlin, Bericht MA 82/2, April 1982

/6/ Wensley, J., Private Communications, Jan 1983

DISCUSSION

Heher: I would like to come back to the comment you made in your paper and your presentation, in which you state that 99% of all error conditions can be detected by the actual expected timing of the computation. I wonder if you could give us some examples of how this error detection operates?

Kopetz: As I mentioned in my presentation, John Wensley reported that in practical experience they had from the AUGUST system, which uses triple modular redundancy, that they found that anything which went wrong in the system first showed up, in the majority of cases, by the timing of the system not being what was expected. You can detect errors by means of time-outs or by finding that one of the three systems gets out of sync with the others. Looking at the timing behaviour of the system and not at the value-domain, such errors can be simply detected. Following on, of course, we can always see that in most cases some kind of change occurs in the value domain, but the point is that the first indication that something has gone wrong in a system is that the timing is not what was expected by the architecture.

LaLive d'Epinay: I would first like to state that I agree with many of the points made by Professor Kopetz. Unfortunately, in the shortness of my presentation, there appeared to be some differences between his concepts and my approach. I think in general we are in total agreement. I think the question of real-time, and of having information of time, as part of the available information, is essential. It is perhaps not that crucial a point and I think that the problem of real-time is just a special case of the problem of data consistency. Certainly one of the goals we have relates to consistency. I meantioned, for instance, that the broadcast principle allows us to have consistently the same information all over the network and with the acknowledgement system you can be certain that all 1100 (or however many nodes you have) have the same information correctly, otherwise none of them have the information. So it is one of the pre-conditions to have atomic actions and data consistency in the systems. The problem of transmitting time is also very important and I totally agree, and as our design shows, we have had physical supportive type information. We can also have complex time and it is absolutely normal to have a type which consists of value and time. For example, we can have two instruments which are measuring the phase angle over a high-voltage line. This cannot be done by having real-time measurements on both sensors then merely combining and taking the differences. We have to take a measurement at one point and than a second maybe 800 km away. We therefore have to know the point-of-time at which these measurements were taken and then we can compute the phase angle between them. I do not think that our concept is any more complex in solving the problems of real-time in the system than any others. I think, in fact, that it has some basic advantages in that it can have consistency of data.

Another point that has been raised, is that the acknowledgement in a broadcast system uses a large amount of the bandwidth of the communications media. This is possible, but not necessarily the case - that is why I described our combination of negative acknowledgement. This means that you need, in fact, one time slot in which every member who has not got the message correctly, sends a negative acknowledgement. If there is too much information you can use the approach in which the negative acknowledgement is a white-noise signal which is sent over the line during the next package of information. So you don't need any bandwidth at all. As to whether the negative acknowledgement is important or not, we have not really discussed this here, and there are clearly installations in which you need it. Just to give an example, you can go anywhere in a power plant and you should be able to plug in a supervising unit and know immediately which parts are working correctly and which are not. So we need negative acknowledgement. The other question which arises is: do we need acknowledgement on the sending or on the receiving end? We don't solve that problem explicitly so we just do it at both ends.

Kopetz: I think in fact we have a very good starting point for discussion as the audience at the workshop consists of many experts on computer control systems. I would like to make a very strong statement - one which is very controversial if you go into certain data processing environments. I feel that there is good reason to have a global, synchronized, real-time facility in a distributed real-time system. It means that every node in the system should have access to global physical time. It certainly appears to me that this is a fundamental issue to be considered in the design of the architecture of future distributed systems. If you look at many of the systems currently being offered for real-time process control, you will find that these do not contain global physical real-time features.

LaLive d'Epinay: In general it appears that most people seem to agree with Professor Kopetz. However, I would like to mention that it is also a matter of cost. In a power plant, a relatively large installation, it is easy to have global physical time. In the power distribution system however which could range all over Europe, and in which you are using a variety of communication systems, it is probably more difficult and costly to provide global real-time. You can have radio stations which transmit synchronized time which can then be used: however, the reliability of such signals depend on atmospheric conditions. I think that the larger the geographical distance which exists between physical processors, then a natural consequence is that the resolution of time which is required will change. So perhaps we should have different time scales for different problems. Maybe we should have different "ticks" and we should be a little bit flexible in specifying the time between ticks as a function of the application which we are involved with. In a particular application you should be able to determine the interval between ticks which can be allowed.

Harrison: I would disagree with Professor Kopetz when he says that data processing applications do not have a need for real-time. Many of them in fact do. The issue is perhaps the resolution that is required and not whether or not it is required. I don't think that people are going to disagree with that and the resolution essentially then becomes a function related to the time constant of the system. When you are talking about a power plant, then I think it is power sequence monitoring functions in which you have relays tripping, etc, and in this case you need to know, with a very fine increment of time, when these events occur. But in other industries, for example in many continuous process control applications, you never run into a case where you really need to differentiate at such a fine detail. I don't think it is an issue of, "Do you need time?" but an issue of, "What resolution do you need?"

Kopetz: It could also be said that in many relatively slow continuous processes the alarm monitoring or the sequence of alarms could require a fine resolution of time, or at least it could be very advantageous to have such a resolution.

Harrison: I would agree to that point, but I am saying that it is a transport mechanism that you are monitoring, in which case alarms physically do not trip faster than, say, one tenth of a second, because that is how fast a conveyor belt, for example, is moving. That is an entirely different time scale than if you are talking about switchgear which can sequence, in a matter of milli-seconds, down through 10 or 20 relays.

Sloman: Is it not more a question of whether time and time-stamping is absolutely fundamental at the lowest levels, or whether it should be put in where it is required? In some cases it may not be vital, in that if your transmission delays in the distributed system are small compared to the rate at which things change, then you might not need to time-stamp every variable that you read, because when you are reading it you are going to get an accurate value since it can't change all that quickly. So you have effectively got an immediate reading.

LaLive d'Epinay: Overall I think it is a question of economics. If you have a temperature which can change, say 1 degree per hour, then perhaps you don't need time-stamps at milli-second intervals. On the other hand, you can have other variables that change very fast and in this case, fine resolution is required.

Kopetz: Does it not make sense then to design real-time into the basic system architecture? In other words, the facility of synchronizing clocks and making available real-time all the way through the system. You can then use it whenever it is required or to whatever grain you need it to be. I feel that in general the basic service provided in a distributed computer system architecture, should give you the facility to access local clocks in each node, and know that the readings you obtain are accurate to a certain grain of time.

Maxwell: In continuous process control you typically want to know how much raw material you have used, and hence, how much profit you have made! If you are measuring such values, you want to integrate them accurately. This is also true in the question of energy consumption. If you are seriously considering comparing energy consumption, you want to integrate the variables monitored on some basis of time. For these reasons you need a time base which is as accurate as you can get it. I think in process control there is no argument about real-time, it is where we start. I always need to keep track of time as accurately as possible.

LaLive d'Epinay: It is strange therefore that most communication systems do not provide time information or are not even deterministic in their time behaviour. I can also state that the best customers of deterministic communication systems are those who have made use of a probabilistic system - ETHERNET for example! These users, perhaps for the first time, begin to feel the need for deterministic time scales once they have used the probabilistic approach.

Maxwell: In our industry we need accurate time scales. We need intervals and we need real-time information - and we required these readings even before we thought about distributed systems! We have always been concerned about how good our real-time clocks were, and how good our software was when it accessed these real-time clocks. We were concerned about how accurately we could measure intervals because this determined how accurately we could measure our material and energy consumption.

LaLive d'Epinay: In the specific problem of distributed systems, this problem becomes very critical. In a local system at least you had logical time. Events occurred in the right sequence and the computer could monitor those sequences. In the case of distributed systems, because they do not automatically have consistent global time, we do not even have logical time.

Kopetz: We essentially seem to agree that there is a need to have synchronized global time, and it appears that it is only a question of the accuracy which is required. The question that must be asked is - "What would users consider to be a reasonable grain of time in a general distributed system architecture?" Do users consider that there exists such a defined grain of time? In our MARS development we have stressed that we wish to develop a general architecture which provides a minimum level of granularity in the time domain. So once again the question is - what should this grain of time be?

LaLive d'Epinay: When you study a particular communication system the question is how much overhead does it require to have the grain of time at a specific resolution? It is a question of the communication protocol. So again it boils down to economics. If a small grain of time was required, would this overload the communication system? Probably there is an asymptotic cost curve involved in attempting to reduce this grain of time.

Kopetz: I think it would depend on the kind of communication system you choose. I would consider that if you use a token-bus with a one megabit speed, then a time resolution of one milli-second is a reasonable resolution to strive for without a particularly high cost.

Imamichi: On our system we have solved the timing problem in another manner. When we need signals which have to be synchronized, we achieve this by ensuring that readings occur at synchronized intervals. This is the case when several computers are working in the same area. Referring to the communication systems, we have produced a system in which there are synchronization signals and these are shown in my paper as existing on the data high-way in which we have incorporated such a synchronization system.

Harrison: I would like to suggest that the issue of absolute time or time invervals should not be fundamental to the architecture. I think that the question of claiming that we should plan real-time into architectures is primarily incorrect. The timing is an implementation problem, and whilst it might be true that your architecture will or will not work over a certain interval of time, I would suggest that it is not correct to try and bound it in absolute terms. If you are talking, say, about 100 milli-seconds then I can suggest applications in which this would fall far short from the real requirements. In modern nuclear scintillation counters, for example, the timing is far more critical than this. So I don't think it is a correct question to ask about architecture. It may well be a proper question to ask about implementation.

Kopetz: I don't fully agree with this statement. I would agree in saying that the absolute grain of time is not a question of architecture, it is a question of implementation for a particular application. But the fact that you do have global time in your system (to a certain granularity) and that you can rely on your protocols to ensure this, then in fact, this is an architectural question.

MacLeod: I would like to raise another point. In computer science the problem of real-time programming has been studied for long and it is recognized that subtle timing problems are very difficult to detect and deal with. The approach which is normally taken is that programs should be proved correct, probably at the same time at which they are developed, and such proof of correctness should be independent of physical time. In fact the current techniques of proving programs correct can only cope with a situation in which no physical time is included. Is this not perhaps an argument against using physical time in the solution space?

Kopetz: I think this is a very good point, but I think that if you prove the logical correctness of parallel processes which are co-operating in the non-real-time domain then this proof will only give you one of properties of your solution space. This is because, in addition to the correctness proof, you would also have to prove that you get your response in the real-time, which is dictated by the environment. I would also assume that having global real-time in a system makes it easier to solve problems in which real-time is not that critical. If you have to order things and if you have real-time available, then it is easier to order things than if you do not have time information. This is true even in a non-real-time application. If one looks at recent publications in the area of transaction processing, one finds that more and more the goals are towards time-stamping. They may, in fact, be logical time-stamped, but basically I believe that physical time-stamping in the system would assist in resolving a lot of the problems.

Rubin: I have been involved with systems which were originally designed to operate with a central clock broadcasting time codes on communication links and having local stations which operated with time codes. In such systems synchronization is indispensible. These systems have shown, in operation, that it is necessary to install local clocks at out-stations and to allow these to be synchronized by a central system. However, the local clocks should be capable of operating on their own when the links fail. So I have no doubt that in a computer network it is possible to do synchronization under normal data transfer by simply transmitting time as data from the reference source. So as to whether this property should be inherent in a communication channel or something which is programmed in when needed, it seems to be a question of economics. Is such a feature worth the money or are there other features which should be built-in which are judged to be more cost effective?

Kopetz: I would like to respond by saying that this is a very good point. In our approach we have come to the conclusion that it is important to have global physical time available at every node in a distributed real-time system. We have also found that there are already implementation techniques available which give this global physical time in a fault-tolerant way. In such a system, even if one of the clocks fails, the system will automatically remove the faulty clock, but continue to operate. These techniques are available and we have considered it necessary to make use of them and to provide global clocks in a fault-tolerant way. Our conclusion is that the added cost is not highly significant.

Van Selm: I would like to take another approach. If we could conceptualise what is happening in each distributed process in terms of the states which are permitted in each node, then one can determine a collection of states in which each processor can exist. You could then have a local real-time facility available in each distributed processor. I would consider then that, perhaps, you could bypass the requirement for global real-time clocks, since the triggering of the operation of one processor from one state to another could be initiated by a signal from another processor. It would not matter then if one processor was out of synchronism with the other because their internal states would be clearly defined. If you had the information relating to exactly what was going on in a process in terms of the state rather than in global time, and if you had local real-time available in each processor, you could initiate transition from one state to another anywhere in the system as long as you knew what the state was.

Kopetz: As far as I can see this would be fine if you stayed within a processor. But if you step outside of a processor and try to get a global view of what is happening in the system, then somehow you have to order the events which are happening in each processor. Once you do this real-time becomes a major consideration. In principle one could make use of logical time, as has been discussed before, but we have found that in order to generate logical time the algorithms are complex and a large amount of traffic in the communication system is required. Further, this traffic, determined by the algorithms, requires reliable transmissions and if one has to make use of an inherently unreliable communication system, most of the algorithms become problematic. In fact, looking at the available algorithms one finds that they all assume totally reliable communication. In practice I think it is probably far more complicated to approach the problem in this way than to provide physical global time as has been suggested.

LaLive d'Epinay: I think if you base any structures on a state model then you find that the state is something which is comprised of more than one node. You have the state of a system as soon as you have this, and you have state transitions in the system and you must be able to order these state transitions - otherwise the normal state model does not work. There is also the other point that as long as things work with local state models then you could probably achieve what Mr van Selm was aiming at, but I think one of the properties of a distributed system is that we attempt to control a complex physical process in which the state models cover several nodes.

Gellie: I think we have summarized what Professor Kopetz was saying. When you are talking about demanding global logical time, even if you do not want global physical time, then this may only be achieved if you have reliable communications. However, global physical time avoids this problem and can operate with unreliable communication schemes and, in fact, be cheaper in terms of network traffic.

Atkins: I would be very interested to know how synchronization can in fact be done? The problem that I see is that you are trying to get a granularity of time which is smaller than the unpredictable delay which occurs in the communication media.

Kopetz: If you have a token-passing network, or something similar, you know how long it will take for one message to get to another node. If, however, we have a system such as ETHERNET, one can still estimate this transmission time provided we know whether a message has been delayed. This presently is not available but future ETHERNET interfaces should include a bit to indicate whether back-off of the message has occurred.

In order to synchronize clocks we do not need to get regular messages because we have local clocks which will only deviate to a small pre-determined amount so that re-synchronization is only required every so often. Therefore if we receive a message with a time-stamp which has been delayed, then we do not use this to synchronize a local clock.

Atkins: There are, however, two problems. The first is that when we introduce a new processor into the network you want to initialize its clock to the global time. The other problem occurs when an individual processor's own clock starts to drift in relation to the others.

Kopetz: These are certainly two separate problems but they are part of the same overall problem of synchronizing and re-synchronizing clocks. When you put in a new processor it will pick up the correct time after a certain period, and it must not transmit a message until this has occurred. There has to be a start-up algorithm.

Kingham: Going right back to the question of the data flow concept of programming. In this approach you have function boxes and triggers. You must have data available to all areas of memory at the same time. It is not quite clear how this fits in with the network concept. In particular, how do you use the network to sequence the events?

LaLive d'Epinay: This is clearly one of the fundamental problems. In practice you cannot, at one instant of time, distribute different kinds of information to all processes. You have to rely on a minimum granularity of time and a serial communication system which must be fast enough to communicate all values which have to be produced within this time span. Within that granularity you can do it purely sequentially or it can be event driven. I think that present systems which work with one megabit transmission rates already are fast enough to cope with large systems. But we feel that in the near future we will be able to improve on this performance. One of the points which I did not mention in my paper, but which is crucial, is that the receiver has to select an address out of all the possible addresses which appear on the communication system. This clearly has to be done extremely fast and there are already custom chips available which can cope with a one megabit system. We have carried out research into new mechanisms which will allow us to identify values which are travelling along the bus at much higher rates - maybe up to 20 megabits. However, the question is still there and there are certainly areas in which we have huge data-base systems in which we cannot communicate all possible data on the single communication systems. This is clearly a limit. In such a case in which one of the nodes has to be a data-base machine, we will have to use other techniques or introduce a hierarchy in which some values are only sent to certain parts of the system. Clearly, although I did not mention it in my paper, we could still have a hierarchical structure.

Further, I think it is important to point out that at the communication system level each node computer will have a separate processor which listens to the communication system. This processor has a table of the names of values which have to be brought into the node. The present system does have a limit of 256 24-bit addresses. The identification is done on-line, and VLSI designs will allow expansion of this. All values are identified by a unique 24-bit identifier. Each node has a table of all the unique identifiers applicable to that node and the system automatically selects those which are necessary for that particular node. The point is essentially that the identification is carried out in hardware, by a look-up table.

A HIERARCHICAL MODEL FOR DISTRIBUTED MULTI-COMPUTER PROCESS-CONTROL SYSTEMS

W. P. Gertenbach

Department of Electronic Engineering, Nuclear Development Corporation, Private Bag X256, Pretoria, South Africa

Abstract. A model for the description of multi-computer process-control systems is described. Its purpose is to act as a reference for the analysis, synthesis or comparison of multi-computer process-control systems. The model involves the decompositioning of the overall system into two-dimensional levels of abstraction which have been selected on the basis of the context within which data is viewed in the system. Several hierarchies of decision-making are used to represent the process-control and other decision-making problems. Functional units are distributed within levels of decision-making in the above-mentioned decision-making hierarchies which are super-imposed on the above-mentioned levels of abstraction and the topology of the plant. Technological features are modelled in terms of a hierarchy of implementation levels and functional units are described in terms of general system traits. The model is used to describe complex distributed computer-control systems, and was extrapolated to the area of artificial intelligence. An important concept, interconnection abstraction, was identified.

Keywords. Distributed computer systems; hierarchical systems; models; process control; computer organization; programming languages; artificial intelligence; computer control; computer interfaces.

INTRODUCTION

Multi-computer process-control systems (MCPC-systems) are complex. Truly complex systems, according to Mesarovic (1970), almost by definition evade complete and detailed descriptions, one of the prerequisites for understanding such systems. Hierarchical decompositioning of the overall system into smaller subsystems is seen partly as a solution to this problem. As will be shown below, several notions of hierarchy and several hierarchies of the same type are necessary in order to describe MCPC-systems. It is also necessary to identify a set of constant traits in terms of which such systems can be described adequately. Klir (1969) has identified such a set of traits which will be used in the above context.

A model for MCPC systems has to comply with the following requirements:

(a) it has to model the entire multi-computer process-control system. It therefore has to cover much more than the Open Systems Interconnection (OSI) model (OSI, 1982). It also has to include the interface to the plant.

(b) in order to ensure a reasonable lifespan for the model, it has to be independent of technology.

A model of MCPC systems is required for the following reasons:

(a) to increase the level of understanding of MCPC systems. Weizenbaum (1979) motivated briefly the use of models in order to improve the understanding of large-scale systems.

(b) to create a logical structure, like the OSI model, which can be used as a reference for the comparison of systems.

(c) to be used as a reference for the structured design of MCPC systems. The model can therefore be used as a framework for achieving such properties as completeness, correctness, consistency, feasibility, testability, re-usability and reliability of the system to be designed.

DECOMPOSITIONING OF THE MCPC SYSTEM

As stated in the introduction, a MCPC system is such a complex system that it cannot be modelled in terms of a single hierarchy. It was found that the following notions of hierarchy were necessary in order to model MCPC systems, viz.:

Levels of Abstraction
Levels of Decision-making/
Organisational levels
Levels of Implementation

The topology of the system also has to be included in the model.

In this section the above concepts and the set of constant traits, as mentioned in the introduction, will be described.

The Set of Constant Traits of the General System

Klir (1969) identified the set of constant traits of the general system which [with some adaptations Gertenbach, (1981, p. 69)] are as follows:

(a) The set of external quantities.
(b) The behaviour of the system.
(c) The set of elements and couplings (EC structure).
(d) The state-transition structure (ST structure).
(e) The activity of the system.

The first two traits, when combined, describe the function of the system. The third and fourth traits describe the organisation of the system. The EC structure is seen as the fixed or permanent part and the ST structure is the variable part or the program of the system. The activity describes the ensemble of the variations in time of some quantities of the system.

The operating principle of a system should be described in terms of its EC structure, ST structure and Activity. For the sake of completeness, the description of complex systems should involve all the above mentioned traits.

Levels of Abstraction

Levels of abstraction (strata) have been defined in mathematical terms by Mesarovic (1970). The decompositioning of a system into levels of abstraction (by abstraction is meant the process by which an idea is stripped from its concrete accompaniments) involves the decompositioning of the system into vertical arranged strata, where, on the lower levels, a more detailed description of the system is given than on the higher levels (Mahmoud, 1977). The set of features in terms of which each stratum is described will be referred to as the context within which subsystems on that stratum operate (Mesarovic, 1970, p. 41).

It was reasoned that information processing is a prime function of a process-control system. This involves the transformation of plant information to a format which is suitable for human absorption and vice versa. For this reason it was decided that a stratification should be based on the context within which data is viewed. A taxonomy (naming scheme) for levels of abstraction which corresponds with the context within which data is viewed on each level of abstraction, is described in a further section. In the OSI reference model (OSI, 1982), the function of the intercomputer communication system is described in the context of the services which any particular level provides to upper levels.

It is important to note that a stratified system description is concerned only with the functions of a system and not the EC structure of its components.

Levels of Decision-Making

A complex system has a complex control or data-processing goal. Mesarovic (1970) proposed that the understanding of this goal can be improved by a decomposition of this goal into a set of subgoals associated with decision units which are ordered within a hierarchy of decision-making levels.

The goal of a decision unit is seen as the objective of its efforts during the execution of its function. A decomposition of the overall goal is thus seen as being related to the decomposition of the overall function of the system. Therefore it can be concluded that a decomposition into levels of functionality and a decomposition into decision-making levels is one and the same thing.

It can therefore be concluded that the OSI reference model, which is based on the principle of function grouping (OSI, 1982, p. 4) is in effect a decision-making hierarchy. This conclusion is important in order to relate the OSI model to the MCPC model to be proposed below.

As will be shown below, a MCPC system can be modelled in terms of multiple decision-making (functional) hierarchies. Again, as in the case of stratification, a decomposition into levels of decision-making does not involve a description of the internal organisation of decision units.

Modularity

Modularity denotes the ability to construct systems with a variety of arbitrary units or modules of standardised dimensions or functions without modification or trimming, or knowledge of the internal organisation of such modules.

It is important to note that modules are sub-systems and it should therefore be possible to characterise them in terms of the five system traits described above, irrespective of their level of implementation.

The subject of modularity as related to MCPC systems has been discussed by Gertenbach (1981); however, for the purpose of this paper it is necessary to review the concept of visibility across module boundaries. Parnas (1972) concluded that the internal operation of a (software) module should not be visible from the outside of the module. It is, however, claimed that the opposite should also be true. Thus, visibility across module boundaries should be limited in both directions.

In the case of hardware implementation technology this is generally true, for example in the case of a LSI module (or component), "knowledge" about its environment, viz. the names of other components, to which the module is to be connected, their behaviour and internal organisation are not "known" within the LSI module. Interconnections between hardware modules are made by means of an external interconnection structure, for example a printed circuit board, a data bus or computer network. It can thus be concluded that the interconnection of hardware modules is abstracted from the modules themselves and is embedded in an external structure such as a printed circuit board or data bus.

It is interesting to review the modularity of software in terms of the above discussion. The following can be concluded:

(a) Generally, software modules know their environment in the sense that the names of modules to which they communicate are embedded within these modules. For example, in an ADA task, the calling task names the task with which it requires a rendezvous (the ADA intertask communication primitive).

(b) Therefore, interconnections between ADA modules cannot be abstracted to a separate interconnection structure.

In the case of single computer systems, this is already an unsatisfactory situation. In multi-computer systems, where inter-task communication is even more difficult, it is felt that the situation is highly undesirable. A possible solution to this problem within the scope of the MCPC reference model will be proposed below.

Levels of Implementation

A model of MCPC systems should, as far as possible, be independent of technology in the sense that technological progress should not affect the structure of the model. However, the role which technology plays in the model should be clarified. In order to do this, a further hierarchy, the hierarchy of methodological levels, or levels of implementation as originally defined by Bell (1971), will be employed. In this hierarchy, according to Hartenstein (1977), each level is characterised by its design activities, the elements which are used for the design, and the complexes synthesized by the design. As stated by Dasgupta (1982), the various levels may come about by a process of construction; a mechanism at level i is constructed on a mechanism at levels i-1. The objective in such cases is to provide some capabilities at level i that were absent at the lower level.

There is obviously a grey area where it may be difficult to decide whether one is dealing with levels of implementation or levels of abstraction. In fact, Dasgupta (1982), in his work on computer design and description languages, made no distinction between the two concepts. It should, however, be emphasised that when one refers to levels of implementation one is concerned with the means or methodology which is employed in order to construct a system as discussed above. On the other hand, when one refers to levels of abstraction, one refers to the context within which the operation of these levels is viewed. Levels of abstraction should be used to describe the functional operation of parts of a system in terms of a specific context. Such a description should be independent of technology or the structure of the system on that level.

In terms of general system traits, it can be stated that in the hierarchy of implementation levels, other than in the other hierarchies discussed thus far, one is concerned with the organisation of the system, therefore also the operating principle on each level of implementation. In fact it should be stated that associated with each level of implementation is a specific operating principle, as will be discussed in more detail below.

THE MCPC REFERENCE MODEL

A global representation (levels still to be identified) of a three-dimensional reference architecture for MCPC systems is shown in Fig. 1. It consists of a base plane (along the X and Y axes) on which the physical distribution of computers, or topology, of the MCPC system is represented. In the example of Fig. 1, three computers P1, P2 and P3, which are connected in a hierarchical topology, are shown.

The MCPC system is further decomposed vertically into levels of abstraction. The Z-axis thus involves a stratification of the system. This stratification, the details of which will be shown later, is done in terms of the context within which data is viewed on each level. The semantics of the data become more abstract as one moves upwards in the hierarchy. Every stratum stretches across the entire horizontal plane and thus the topology of the system is mapped onto every stratum. Conversely, a particular level exists (in a logical sense) within all processors in the system.

The process-control decision problem is incorporated in the model in terms of a decision-making hierarchy which is superimposed on the topology plane and is distributed across various strata according to the context in terms of which data in that decision unit (functional unit) is viewed. This is shown graphically in Fig. 2 where functional units are distributed horizontally according to the topology of the MCPC system and vertically according to the level of abstraction in terms of which their data-processing function is viewed.

Nothing has been said up to this stage about implementation in the above description. In fact, the model as described so far is independent of implementation. The hierarchy of implementation levels should be used in order to describe the implementation technology.

It can be stated that in the above model a decision unit (functional unit) is described in terms of the following attributes:

(a) Its level of abstraction: this describes the context of its data-processing function.

(b) Its level of decision-making: this describes its locality within a decision-making hierarchy.

(c) Its level of implementation: this describes the technology which is used in its construction.

(d) System traits such as function, organisation and activities.

As stated in the introduction, the model should take into account communications between computers and also between computers and the process under control. This is illustrated in terms of Fig. 3, which can be seen as a side view of Fig. 1 in the direction of the X-axis and including only two computers. In this figure the higher levels are supported by a structure consisting of two legs. One leg, the process leg (PL), is supported by the plant itself, for direct control of that section of the plant which is under the control of the relevant computer. The other leg, the communications leg (CL), is supported by the communications media for control and communication with that section of the plant which is remote from the first computer. The upper levels communicate with the operator. The solid lines, showing intercomputer communications in Fig. 1, now become the peer to peer communications as shown in Fig. 3.

In accordance with requirements given above for modularity of functional units, it is required that environmental characteristics should not be embedded within such functional units. Thus Fig. 3 proposes that somewhere in the hierarchy there should exist a level (level N + 1) where functional units do not know whether they get plant information directly from their host process-leg or from a remote process-leg in another computer. The N-layer should also provide inter-process communication services which are accessed in a similar fashion whether the communicating processes are in the same or different computers.

In the following sections, proposals are made for the names of various levels of the hierarchies discussed thus far. At this stage it is necessary to clarify where the OSI reference model fits into the above model. Since the OSI reference model is a functional decomposition of the interconnection requirements between communicating computers, it is seen as one of the several decision-making (functional) hierarchies which are superimposed onto the three-dimensional model described below.

LEVELS OF DECISION-MAKING AND DECISION-MAKING HIERARCHIES

As discussed in the previous section, the process-control decision problem should be decomposed into levels of decision-making and functional units which are distributed across levels of abstraction and the topology of the plant. The exact structure of this decision-making hierarchy for a specific plant will depend on the goals which have been set in order to control that specific plant. One possible decompositioning can be as follows:

Level 1: (lowest) Selecting Function. Input/Output functional (decision) units select specific plant parameters for surveillance or control.

Level 2: Regulating Function. Decision units decide on the best outputs in response to specific setpoint and measurement values.

Level 3: Supervising Function. Decision units optimise setpoints and perform general supervisory control functions.

Level 4: Self-Organising Function. Decision units improve or adapt the functioning of lower levels and may perform typical AI functions (Artificial Intelligence).

Level 5: Ergonomising Function. Decision units adapt lower-level information to human requirements and vice versa.

In addition to the decision-making hierarchy which stems from the process-control decision problem, several other hierarchies which may co-exist within MCPC systems can be identified as follows:

(a) A decision-making hierarchy for the management of alarms.

(b) A decision-making hierarchy for global system management (allocation of global resources, configuring, etc.).

(c) A decision-making hierarchy for the management of redundancy. (It disables outputs to plant as a result of malfunctioning in one computer and enables outputs from another.)

(d) A decision-making hierarchy for self-testing and repairing.

Every one of the above decision-making hierarchies involves the decomposition of some decision-making problem into various levels of decision-making and functional units which operate on these levels. The functional units are distributed within the topology of the plant and on various levels of abstraction.

FORMULATION OF THE SET OF IMPLEMENTATION LEVELS

Bell (1971) and Hartenstein (1977) identified a set of seven levels of implementation. Dasgupta (1982) introduced the concept of exo-architecture and endo-architecture in this context. For the MCPC system environment, the following set of levels are proposed, viz:

The component level (lowest)
The circuit-design level
The logic-design level
The machine-organisation level
The machine-architecture level
The abstract-machine level (highest)

Before a brief discussion of each of the above levels of implementation is presented, a few general principles of levels of implementation will be discussed.

The Operating Principle

Some reflection on how systems are implemented will reveal that on all levels of implementation, there may be, to some extent, a subdivision of the implementation level into two parts. The first part is concerned with the flow of data and the second part is concerned with the flow of control which is required to co-ordinate the flow of data. The data involved is closely related to the function or purpose of the system, whereas the flow of control is concerned mainly with the internal organisation of the specific level. The above identification of the dual nature of hardware-orientated implementation levels was made by Hartenstein (1977, p. X111). De Marco (1978) developed a technique of data-flow diagrams which can be used to represent the data-flow structure on the abstract-machine (software) level of implementation. Gertenbach (1981) used a structuring technique to describe on the same diagram the flow of both data and control in real-time software-systems. As a result of this, one can conclude that on software levels of implementation, as on hardware levels of implementation, a structure of elements (data processors) and couplings (data flow and control flow), which represent partly the organisation of the software system (in a general systems context) can be identified. Further reflection on this EC structure will reveal that on software implementation levels, that part of the general-purpose operating system which provides facilities for inter-process communication, resource management and scheduling, plays an important role in keeping this EC structure together. It provides the mechanisms for the flow of control and data between elements and, in general, manages operations on this level. The concept of an operating system can also be identified on other levels of implementation, as will be discussed below. In general terms, it could be called the operating principle of an implementation level. On some levels it is abstracted to a separate entity as in the case of software implementations; in other cases it is merely the description of the organisation of that implementation level. For example, on the logic-design implementation level, the principle of the state machine can be seen as the general-purpose operating principle. It can thus be concluded that the notion of an operating system, as known on software levels of implementation, is not unique to that level of implementation but also applies to other levels of implementation.

Thus, in terms of the MCPC reference model, the concept of an Operating System (on software levels of implementation) is expanded to the concept of the Operating Principle, which on all levels of implementation partly describes the organisation of those levels.

The Communication Structure

Although topology of the MCPC system is modelled on the base plane of Fig. 1, it should be emphasised that the communication structure between elements of an EC structure is an implementation-orientated concept which should be reviewed on all levels of implementation.

The communication structure between processes can be characterized in terms of the following structures:

One to one
Many to one
Many to many
One to many

It is interesting to note that the ADA rendezvous offers a one-to-one and many-to-many communication structure between data processing elements only.

A brief description of each level of

implementation is as follows:

The Component Level

Complexes which are constructed on this level are electronic components such as resistors, capacitors, transistors, etc.

The Circuit-Design Level

The complexes which are constructed on this level from elements on the lower level are logic circuits such as AND gates, OR gates, etc. Interconnections are abstracted onto, for example, a printed circuit board. The operating principles of electronic circuits are applicable to this level.

The Logic-Design Level

The complexes which are constructed on this level from elements on the lower level, are combinational, sequential and data-storage elements such as latches, flip-flops, etc. Interconnections are abstracted as on the circuit-design level. The operating principles of, for example, Boolean logic and the state machine, are applicable to this level.

The Machine-Organisation Level

The complexes which are constructed on this level from elements on the lower level are real computing machines. This level is also referred to as the register-transfer level. It is sometimes decomposed further into sublevels which have been referred to as the micro-architecture and endo-architecture (Dasgupta, 1982).

Interconnections are abstracted as on the circuit-design level. The operating principle may be that of the Von Neumann machine, object-orientated machines (Zeigler, 1981), hierarchical function distribution (Giloi, 1982), array processors, etc.

The Machine-Architecture Level

The machine architecture is that logical structure and capabilities which are visible to the machine-language programmer.

On this level the architectural properties of computers (instruction set, etc.) which are available to the programmer are constructed in a logical rather than a physical sense from lower-level components.

Interconnections are abstracted in a logical sense. The operating principle on this level is an abstraction of the operating principle of the machine-organisation level. It can be concluded that the machine architecture is an abstraction of the machine-organisation level.

The Abstract-Machine Level

The solution of a problem on a computer involves the utilisation of a set of basic operations and data types and an algorithm in order to manipulate these operations. These basic operations and data types can be used to define a fictitious computer which is ideally suited to the problem at hand. This computer is called an abstract machine, and the algorithm is coded in one language or another for this machine. In order to run the program on a real computer, the abstract machine must be implemented on that computer (Poole, 1975, p. 193).

The abstract-machine level can be viewed as consisting of several sub-levels; thus a hierarchy of abstract machines can be visualised. Alternatively this hierarchy could be seen as an extension of the hierarchy of implementation levels. The lowest-level machine of this hierarchy is the assembler machine. Machines near the top of the hierarchy should be compatible with the requirements of the application and should provide data types, operators and languages for the coding of the relevant problem-orientated algorithms.

For the purpose of this paper two abstract machines are important, the first being the general-purpose abstract machine (GPAM) and the second, the process-control abstract-machine (PCAM).

The General-Purpose Abstract Machine (GPAM)

On the GPAM level, elements of an assembler machine are used to build a GPAM.

Implementations of, for example FORTRAN, PASCAL and ADA are general-purpose abstract machines. From the discussion above and as discussed below in the section on the PCAM, it can be concluded that the GPAM should offer facilities to the PCAM for the structuring of the global PCAM algorithm or decision-making hierarchy of functional units in terms of a structure of elements and couplings. This involves:

(a) GPAM processes (elements) which can assume the rol of data processors. These processes should be characterisable by general system traits as discussed above.

(b) Interconnection facilities between GPAM processes, which provide intercommunication patterns to co-operating processes (even in remote computers) as described above, are required.

(c) Facilities for the abstraction of interconnections between GPAM elements to a structure (an interconnecting structure) which are external to GPAM elements, are required. Thus GPAM elements should not know the names of elements to which they communicate at the time when they are constructed.

The ADA task concept is seen as a possible candidate to act as a GPAM process; however the ADA language will have to be extended in order to cater for interconnection abstraction. This can be made possible by means of "exit" and "link A and B to C and D" language constructs as illustrated in the appendix and as discussed by Gertenbach (1981, p. 160).

The Process-Control Abstract Machine (PCAM)

On the PCAM level, elements of the GPAM are used to build a process-control software system.

According to Halstead (1979) a program consists of an ordered string of operators and operands and nothing else. Thus, in order to specify the PCAM, it is necessary to identify the set of operators and operands which are required by the PCAM user.

The set of operands of the PCAM language should be orientated towards the process-control problem. They should therefore be a set of general-purpose algorithms which can be interconnected in order to control the plant. The algorithm should be a software component (Johnson, 1981) which can be bought from a catalogue with specifications of precision, normalised speed and space requirements, etc. The PCAM operands can be implemented in terms of GPAM tasks with modifications as discussed above.

The set of operators for the PCAM can now be defined as those operators which are necessary in order to enable control or data flow between PCAM operands. Thus the PCAM operators can be implemented in terms of those facilities provided for interconnection abstraction on the GPAM level. Programming on the PCAM level thus becomes similar to programming on analogue computers.

The above concept is not entirely new; it has actually been implemented on, for example, programmable controllers (in terms of ladder diagrams), Siemens Teleperm-M (Borsi, 1979) systems and Leeds and Northrup MAX-1 (Leeds and Northrup, 1980) systems. The structuring of the process-control system, as it is called in the Siemens Teleperm-M system, is nothing other than programming on the PCAM level.

In Gertenbach (1981), various other proposals, such as the concept of the abstraction of interconnection configuration and a graphic PCAM language, are made. The latter is possible because PCAM operands lend themselves to graphic representation, for example, in the case of programmable controllers.

Advanced Abstract Machines

As discussed above, the GPAM level can be viewed as a level at which software components are constructed. Thus the GPAM level can also be called a software-component level. The programming process on the PCAM is analogous to design procedures on the circuit-design level discussed above. The PCAM can thus also be seen as the software circuit-design level. Thus if one may extrapolate in this case, a software logic- design level, a software machine-organisation level and a software machine-architecture level can be postulated. Perhaps the instruction set of such a software machine architecture can be the means in terms of which the level of concepts which is required for (Artificial) Intelligence (AI), as discussed in the next section, can be implemented.

FORMULATION OF THE SET OF STRATA

It is the purpose of this section to propose a set of strata which is applicable to the process-control environment, and also to describe briefly the functions of decision units on each stratum. As discussed above, it was decided to define a taxonomy (naming scheme) for levels of abstraction in terms of the context within which data is viewed, and which takes into account the increased degree of abstraction of data as it moves upwards from the plant to the operator. This led to identifiers such as the signal level, the bit level, the frame level, the packet level, the message level, the decision level, and the level of concepts as shown in Fig. 4.

In this taxonomy the semantics of information increases as one moves higher up in the hierarchy. The functions of the levels of Fig. 4 are discussed in detail by Gertenbach (1981) and are discussed briefly below.

The Physically Controlled Medium

The physically controlled medium in the communications leg is the actual communications medium, the function of which is to transfer data between data stations. It transfers information between data stations under control of stimuli which it receives from the signal level. For the purpose of this model the physically controlled medium in the communications leg is the actual data-transport medium. It excludes items such as line couplers or data-circuit equipment (DCE). It is not part of the OSI model, though it does form part of the ECMA-80 (1982) specification for local-area network coaxial-cable systems.

The physically controlled medium in the process leg is the actual plant under control. It accepts raw materials, and under the direction of stimuli which it receives from the process computer via the process interface and actuators, it produces a product.

The Signal Level

Data received by the signal level from the bit level is viewed within the context of physical signals which represent information which must be transported down to the communications medium or plant.

The signal level of abstraction contains those functional units of the process-control or other decision-making hierarchies which transform the information received from the bit level to physical signals and vice versa. These physical signals may be optical, acoustical or electronic in the communications leg and thermodynamic, mechanical, etc., in the process leg.

In the communications leg, the signal level involves various devices which are normally referred to as data-circuit equipment (DCE) or line couplers, which, under the direction of the bit level, perform the task of establishing, maintaining and releasing the connection to the communications medium. It performs a semantic and syntactic transform on data from data bits to electronic signals. Its functions are not described in the OSI model, though its logical representation, which is called the physical media connection (OSI, 1982, 4.7.4), is described. Its functions are similar to those lower-level ECMA-81 (1982) functions (for example the recovery of data timing, transmit, recieve, carrier sense, signal-error detection, etc.) which can be considered within the context of signal handling.

Signal-level functions in the communications leg are usually implemented on the circuit-design implementation level

Signal-level functions in the process leg are implemented by means of signal conditioning electronics, transducers and actuators which transform data from/to the bit level to/from the plant as required by the plant/bit level.

The Bit Level

Data received by the bit level from the frame level is viewed in the context of bits of information which must be transported down to the signal level.

The bit level of abstraction contains those functional units of the process-control or other decision-making hierarchies which transform the information received from the frame level to bits and vice versa.

Bits are the smallest quanta of information which are transmitted by the signal level.

Functional units in the communications leg perform those functions which are necessary to activate, maintain and de-activate the physical connection for bit transmission between data-link entities. The bit level therefore contains all the functional units of the OSI physical layer and those functional units of the ECMA-81 (1982) physical layer (for example encoding, decoding, preamble generation/removal, etc.) which can be considered as within the context of bit handling.

Functional units in the process leg perform those functions which are normally performed in a conventional process interface. They provide mechanical, electrical and procedural facilities for interfacing the process computer to process actuators and transducers. The bit level in the process leg is usually sub-divided into several sub-levels which may or may not exist over the entire distributed system topology. In Fig. 5 such a set of sub-levels is proposed. The lowest sub-level may contain functional units which may override or de-activate process-computer outputs to the plant in reaction to a protection system (for example in the case of nuclear-reactor control). In several process interfacing systems, for example CAMAC (CAMAC, 1975) or MEDIA (Harshall, 1977), the equipment between the Input/Output (I/O) bus of the computer and the signals to the plant may be organised into several sub-levels with various intermediate data buses and intermediate controllers. The need for such intermediate buses was discussed by Gertenbach (1981, p. 209).

Bit-level functions in both the communications leg and the process leg are implemented, usually, on the logic-design level or even higher levels of implementation in the case of microprocessor based interfaces.

The Frame Level

Data received by the frame level from the packet level is viewed within the context of an optimum collection of bits which should be transmitted reliably from a data source to a data sink along a route which does not involve intermediate relay stations (OSI, 1982).

The frame level of abstraction contains those functional units of the process-control or other decision-making hierarchies which transform the information received from the packet level to frames and vice versa.

Frames are sets of bits which can be

transported optimally (in terms of cost and reliability) between data sources and data sinks via data links which do not contain relays.

Functional units in the communication leg of the frame level perform the functions of the OSI data-link layer and the ECMA-82 (1982) link layer.

Functional units in the process leg perform functions which can be referred to as interface-bus control. The frame level relieves the packet level in the process leg from detailed I/O operations and the handling of the process interface flags and interrupts. It receives data from the packet level and "frames" it with control signals in order to ensure its delivery to the correct destination (transducer/ actuator). In the upward direction it responds to the process-interface signals and assembles "packets" of data before delivering them to the packet level. It also controls the flow of information to the packet level as in the communications leg. The functional unit which performs the frame-level function in the process leg is usually referred to as an input/output driver (Gertenbach, 1981, p. 303).

Frame-level functions in both the communications leg and the process leg are usually implemented on the abstract-machine level and lately also on the machine-organisation level.

The Packet Level

Data received by the packet level from the message level is viewed within the context of indivisible data units which must be routed via one or more intermediate relay stations to their destination processor or down the process leg to a specific transducer.

The packet level of abstraction contains those functional units of the process-control or other decision-making hierarchies which transform the information received from the message level to packets and vice versa.

Packets are quanta of information which are padded with control data which will ensure their end-to-end delivery via a network or process interface where the former may contain relay stations.

Functional units on the communications leg of the packet level perform the functions of the OSI network layer. It should, however, be noted that in distributed process-control systems where the communications medium is passive and thus does not contain relay stations, packet-level functions are redundant. In such cases packet-level functions do not have to be invoked.

Functional units in the process leg perform functions which can be referred to as the typical operating-system function of input/output control. They provide a service to the upper levels for the transport of data down the process leg. The selection of the appropriate Input/Output driver from, for example, a device reference table which is accessed by a logical channel number, as in some operating systems, is considered to be analogous to the communications leg virtual- circuit concept.

Packet-level functions in both the communications leg and the process leg are implemented, usually, on the abstract-machine level.

The Message Level

Data received by the message level from one or more data sources on the decision level is viewed in the context of indivisible messages which must be transported to one or more destinations which may be in the decision level or down in the process leg. These indivisible messages are the outcome of decisions made by the decision level (for example, the temperature of the furnace is 50 °C, the valve must be closed, etc).

The message level of abstraction contains those functional units of the process-control or other decision-making hierarchies which transform information received by the message level into segmented messages and vice versa. It also provides the facilities for communication between functional units on the decision level. Segmented messages are those quanta of information which permit the proper time-sharing of the underlying communications medium or the proper utilisation of process-leg facilities.

The message level is the first level in the hierarchy which is not subdivided into a process leg and a communications leg. Thus its main function is to render to higher levels a uniform set of services which are independent of whether its peer entities are located in the host or remote computer or whether the message is to be transmitted to the local or remote process leg. The message level thus provides a uniform interface to its upper level but speaks to the process in its language and to the communications leg in its language. The segmentation of messages for the process leg is not necessary, provided that the process leg can handle large blocks of data and single I/O statements concurrently. Owing to the parallel and relatively fast rate of data transmission in the process leg, this is usually the case. The message level, as far as the process leg is concerned, thus has no function other than the provision of uniformity, as discussed above.

Apart from the above functions related to the process leg, the functional units on the message level perform the functions of the OSI transport layer.

Message-level functions are implemented, usually, on the abstract-machine level.

The Decision Level

Data received by the decision level from the level of concepts is viewed in the context of concepts which must be interpreted in terms of decisions or messages which are to be transported between decision or functional units on the decision level.

The decision level of abstraction contains those functional units of the process-control or other decision-making hierarchies which perform the application-orientated decision-making function. These functional units make decisions or generate outputs (messages) which have to be conveyed to other decision-making units within the EC structure of the decision-making level. Examples are regulatory control and supervisory control units, as discussed above.

Messages or decisions are therefore the quanta of information which decision units (functional units) may wish to communicate between themselves.

The decision level may be implemented in terms of the abstract-machine level or preferably the process-control abstract machine (PCAM) level. In the latter case it will require an interconnection structure in order to control the dialogue between decision units. This interconnection structure should be made possible by the general-purpose abstract machine (GPAM) by means of the technique of interconnection abstraction as discussed in a previous section. It was also shown in that section how interconnection abstraction can be achieved by ADA entry statements, entry calls, accept statements and two new language constructs, an exit statement and a link statement. Gertenbach (1981, p. 240) suggested that the facilities provided by an interconnection abstraction structure and the OSI session layer are similar. Therefore it can be concluded that the OSI session layer and those GPAM facilities which provide for interconnection abstraction (see package INTERCONNECTION STRUCTURE in the Appendix) are one and the same thing. Thus the OSI-session layer should be seen as part of the MCPC system implementation technique which is responsible for the interconnection of GPAM processors.

The Level of Concepts

The concept level of abstraction contains those functional units which are responsible for the development or generation of concepts which are to be transmitted to the decision making level.

Concepts are seen as those units of information, the ensemble of which forms the knowledge of the system which houses these concepts.

The level of concepts may be implemented in terms of a human being or an AI machine as discussed in the section on advanced abstract machines.

If the level of concepts is implemented in terms of a human operator, then the interface between the level of concepts and the level of decisions is the conventional man-machine interface. The information which flows through this interface consists of concepts which are generated/interpreted by the operator. However, this interface requires a syntax conversion in order to ensure mutual understanding between man and machine. This is a similar function as offered by the OSI-presentation layer. Syntax conversion may however, also be necessary in order to ensure that decision units "understand" each other. In this case a presentation layer should also be inserted on the level of decisions between decision units and their underlying interconnection structure (session layer).

When an operator communicates with a process plant, he may wish to do mainly one of three things: select a process parameter for display, control a parameter or examine the status of a parameter. These functions are seen as the user-specific application layer functions as identified in the OSI model.

THE MAPPING OF TYPICAL DISTRIBUTED-SYSTEM ARCHITECTURES ONTO THE MCPC REFERENCE MODEL

In Fig. 6 a hypothetical architecture is shown which is a combination of several architectural principles of industrial systems. This hypothetical architecture involves the following:

(a) A single global highway system with special-purpose data stations 1 to 6 connected to it. The Teleperm M (Borsi, 1979), PCI-4000 (Gitelson, 1981) and Contronic P (Hartmann and Braun, 1981) systems conform with this architecture.

(b) An extention of (a) by means of a "gateway" between a local highway and the above-mentioned global highway no. 1. The Yew series 80 (Yokagawa, 1981) and Modicon Modbus/Modway (Allen, 1981) systems involve such an arrangement.

(c) An extension of (a) by means of multiple highways and possibly a centralised highway traffic director (data station 9) for example TDC-2000 (Volpe, 1981) and the

PROWAY system (PROWAY, 1981).

(d) Store-and-forward data stations as illustrated in terms of data stations 12 and 13. Examples are the Hewlett Packard DS-1000 (Hewlett Packard, 1978) and the CERN SPS control system (Crowley-Milling, 1973).

A stratification of a part of the hypothetical architecture of Fig. 6 is shown in Fig. 7. (Data-station numbers correspond in order to illustrate the mapping of Fig. 6 onto Fig. 7.)

CONCLUSIONS

The main architectural features of the MCPC reference model are as follows:

(a) Vertical decompositioning of the system into levels of abstraction which have been defined according to the context within which data is viewed in the system.

(b) A horizontal plane within which the topology of computers is pictured.

(c) Decompositioning of process-control and other decision-making problems firstly into separate hierarchies, secondly into several levels of decision-making, and thirdly into decision units (functional units) which are distributed within the topology of the plant and the above-mentioned levels of abstraction.

(d) A hierarchy of implementation levels has been described which should cater for technological advances.

A specific operating principle is associated with <u>every</u> level of implementation. The principle of interconnection abstraction should be implemented in order to comply with requirements of modularity. The implementation of a session layer in real-time process-control systems can be done in terms of the technique of interconnection abstraction, as illustrated in the Appendix.

(e) Functional units implemented in terms of any level of implementation, distributed within any level of abstraction or decision-making or anywhere within the topology of the plant, should be described in terms of the five general system traits.

The mapping of the architectural features of some systems onto the model has been illustrated. Some reflection will reveal that the model can be used to describe the organisation of a system of distributed robots (if the level of concepts is implemented on the level of implementation of the AI abstract machine) as well as a simple three-term controller (where many of the levels as discussed, will not be invoked).

Detailed investigations into the enhancement of ADA to cater for interconnection abstraction should still be done. The standardisation of a graphics process-conrol language on the PCAM level of implementation should be investigated. The implementation of AI in terms of a software-machine architecture could be researched.

REFERENCES

Allen, B.S. (1981). Data Highway Links Control Equipment of any number of Different Manufacturers. <u>Control Engineering</u>. July 1981.

Bell, C.G. and Newell, A. (1971). <u>Computer Structures: Readings and Examples</u>. Mc Graw-Hill.

Borsi, L. (1979). Teleperm M - A Microprocessor-based Automation System with Decentralized Structure. <u>Siemens Power Engineering</u> No. 8, 249 - 253.

Crowley Milling, M.C. (1973). The CERN SPS Multi-computer and CAMAC Real-Time Control System. <u>Proc. 1st. Intern. Symp. on CAMAC</u>, 121 - 128.

Dasgupta, S. (1982). Computer Design and Description Languages. <u>Computer Advances</u>, 91 - 154.

De Marco, T. (1978). <u>Structured Analysis and System Specifications</u>, Yourdon, New York.

ECMA-80, (1982). Local area networks coaxial cable system.' <u>ECMA Standard</u>. Sept. 1982.

ECMA-81, (1982). Local area networks physical layer. <u>ECMA Standard</u>. Sept. 1982.

ECMA-82, (1982). Local area networks link layer. <u>ECMA Standard</u>. Sept. 1982.

Gertenbach, W.P. (1981). An investigation into the organisation and structured design of multi-computer process-control sytems. <u>PhD Thesis</u>. University of Natal. Durban. South Africa.

Giloi, W.K. (1982). Multicomputer architectures - the development of the eighties. <u>Proc. 2nd. Int. Symp. on Real Time Data</u>. Versailles 3 - 5 Nov. 1982.

Gitelson, J. (1981). Configuration of Distributed Control Systems Incorporating both Analogue and Sequential Control. <u>SACAC Symp. on Distributed Systems</u>. Durban. 23 - 24 April 1981.

Halstead, M.H. (1979). Advances in Software Science. <u>Computer Advances</u>. 1979. 119 - 172.

Harshall, J.R. et al. (1975). Media, a Continuous Digital Process Control System. <u>ISA Transactions</u>, Vol. 12, no. 3, 281 - 285.

Hartenstein, R.W. (1977). <u>Fundamentals of Structured Hardware Design</u>. North Holland. 323 p.

Hartmann & Braun. (1981). Contronic P. <u>Technical Data</u>.

Hewlett Packard, (1978). Distributed Systems DS-1000. <u>Technical Manual</u>.

Johnson, R.C. (1981). Special report: Ada the Ultimate language. Electronics. Feb 10, 1981. 127 - 132.

Klir, G.J. (1969). An Approach to General Systems Theory. Van Nostrand. 323 p.

Mahmoud, M.S. (1977). Multilevel System Control Applications: A Survey. IEEE Trans. on Systems, Man and Cybernetics. Vol SMC-7 no. 3. March 1977.

Mesarovic, M.D., Macko, D., Takahara, Y. (1970). Theory of Hierarchical, Multilevel Systems. Academic Press.

OSI. (1982). CCITT Proposed Draft Recommendation - Reference Model of Open Systems Interconnection for CCITT Applications. Sept. 1982.

Parnas, D.L. (1972). On the criteria to be used in Decomposing systems into Modules. Communications of the ACM. Dec. 1972. Vol. 15, no. 12, 1053 - 1058.

Poole, P.C. (1975). Portability and Adaptability. Lecture Notes in Computer Science. No. 30. Ed. Goos. Springer Verlag.

Proway. (1981). Process data Highway for distributed process-control system. IEC subcommittee 65A. Draft Specification.

Volpe, A.A. (1981). Architecture for an Uninterrupted Automatic Control System. SACAC Symp. on Distributed Systems. Durban. 23 - 24 April 1981.

Weizenbaum, J. (1979). Human authority, responsibility and accountability in large-scale real-time systems. Proc. first European Symp. on Real-Time data handling and process control. Berlin (West). 23 - 25 Oct. 1979.

Yokagawa. (1981). Yew series 80 process control system. Technical Manual.

Ziegler, S. (1981). ADA for the INTEL 432 Microcomputer. Computer. June 1981. 47 - 56.

APPENDIX

The following is an example of the implementation of interconnection abstraction in ADA with two new language constructs "link" and "exit" which, together with the standard "accept" and "entry" statements, interconnect two tasks A and B.

```
package INTERCONNECTION_STRUCTURE is
        task A is separate;
        task B is separate;
end     INTERCONNECTION_STRUCTURE;
```

```
package body INTERCONNECTION_STRUCTURE is
begin
        link A exit OUT to B entry IN;
        link B exit OUT to A entry IN;
end     INTERCONNECTION_STRUCTURE;
```

The following codes illustrates the coding of task A.

```
with     INTERCONNECTION_STRUCTURE;
separate task A is
         entry IN(P : in INTEGER);
         exit OUT(Q : out INTEGER);
end A;
task body A is
         STORE : INTEGER;
         Q : INTEGER;
begin
         Q : = 0;
         OUT (Q);
         .
         .
         .
         accept IN (P : in INTEGER);
         STORE : = P;
         end IN;
         .
         .
         .
end A;
```

The following coding illustrates the coding of task B.

```
with     INTERCONNECTION_STRUCTURE;
separate task B is
         entry IN (R : in INTEGER);
         exit OUT (S : out INTEGER);
end B;

task body B is
         SAVE : INTEGER;
         S : INTEGER;
begin
         accept IN (R : in INTEGER) do;
         SAVE : = R;
         end IN;
         .
         .
         .
         S : = 0
         OUT(S);
         .
         .
         .
end B;
```

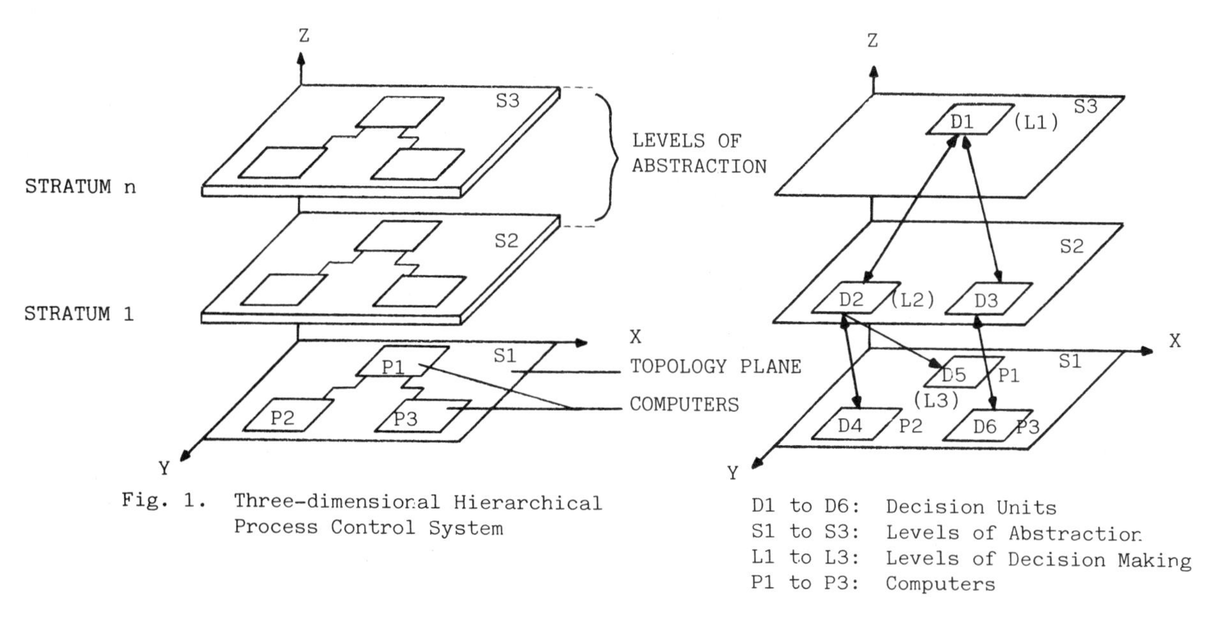

Fig. 1. Three-dimensional Hierarchical Process Control System

D1 to D6: Decision Units
S1 to S3: Levels of Abstraction
L1 to L3: Levels of Decision Making
P1 to P3: Computers

Fig. 2. A Decision-Making Hierarchy Superimposed on the Topology Plane and Three Levels of Abstraction

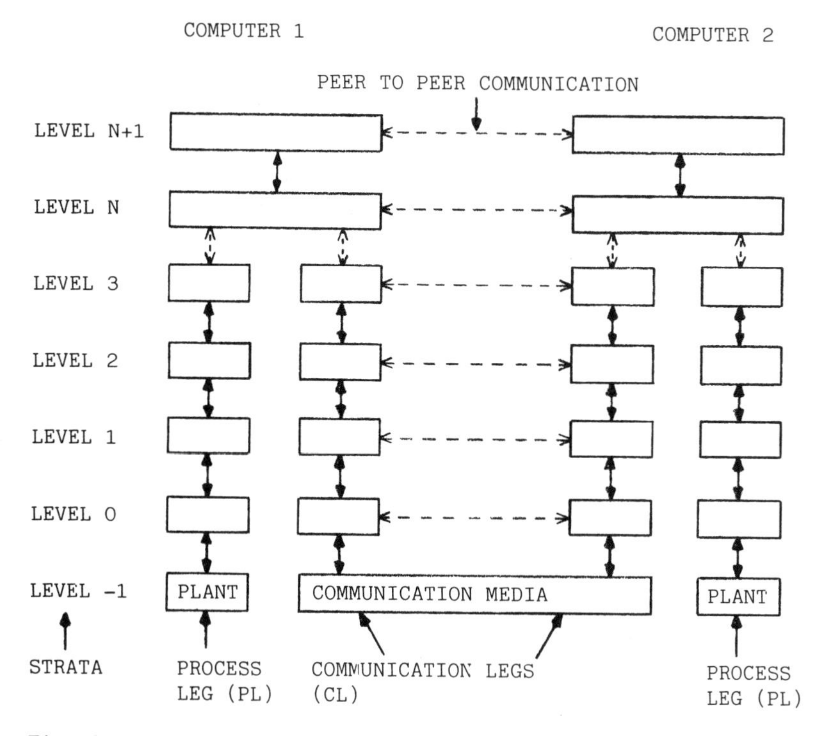

Fig. 3. A Side View of Fig. 1 showing both Process Control and Inter-Computer Communication (only two computers shown)

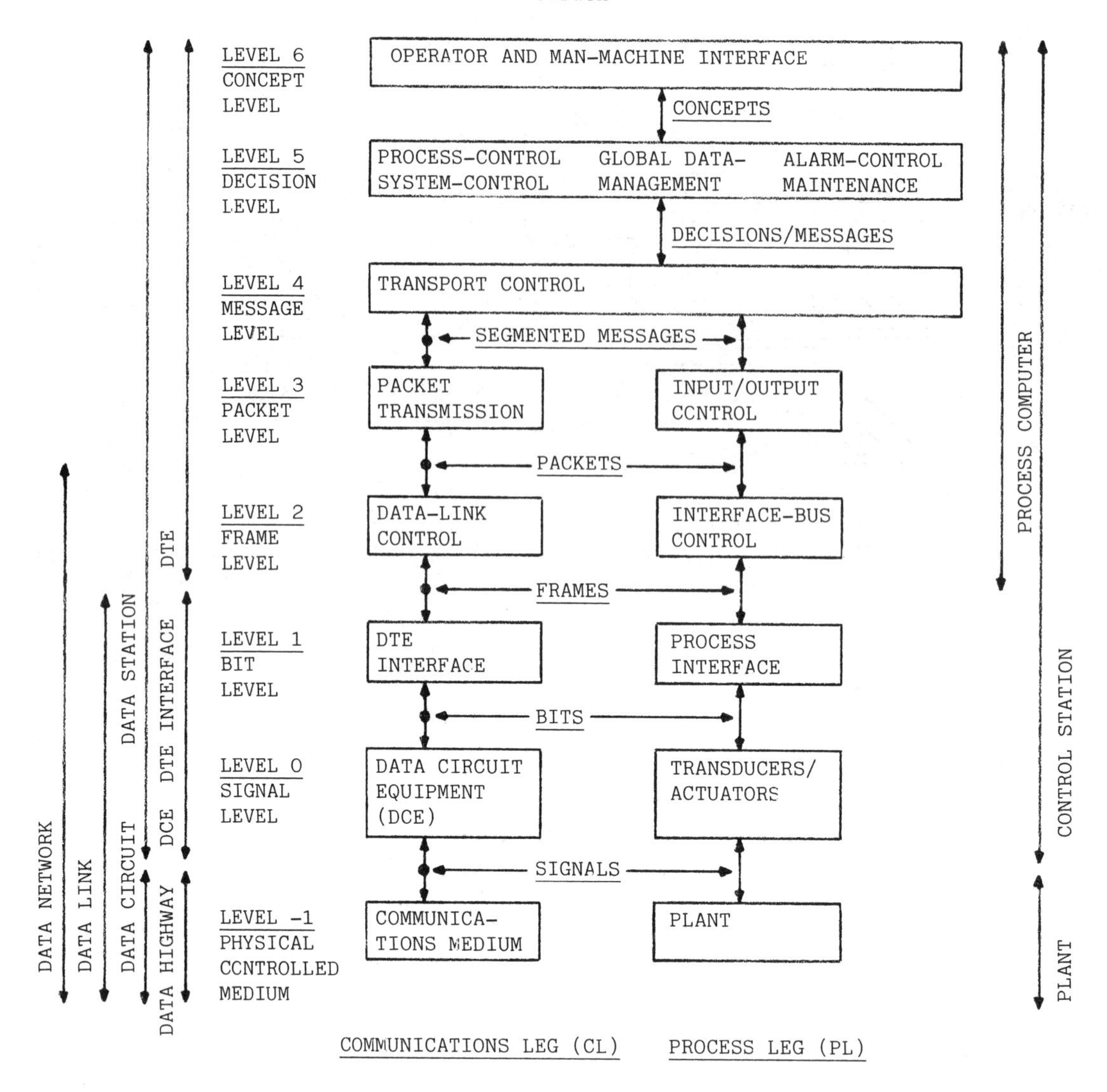

Fig. 4. The Strata of the MCPC-Reference Model

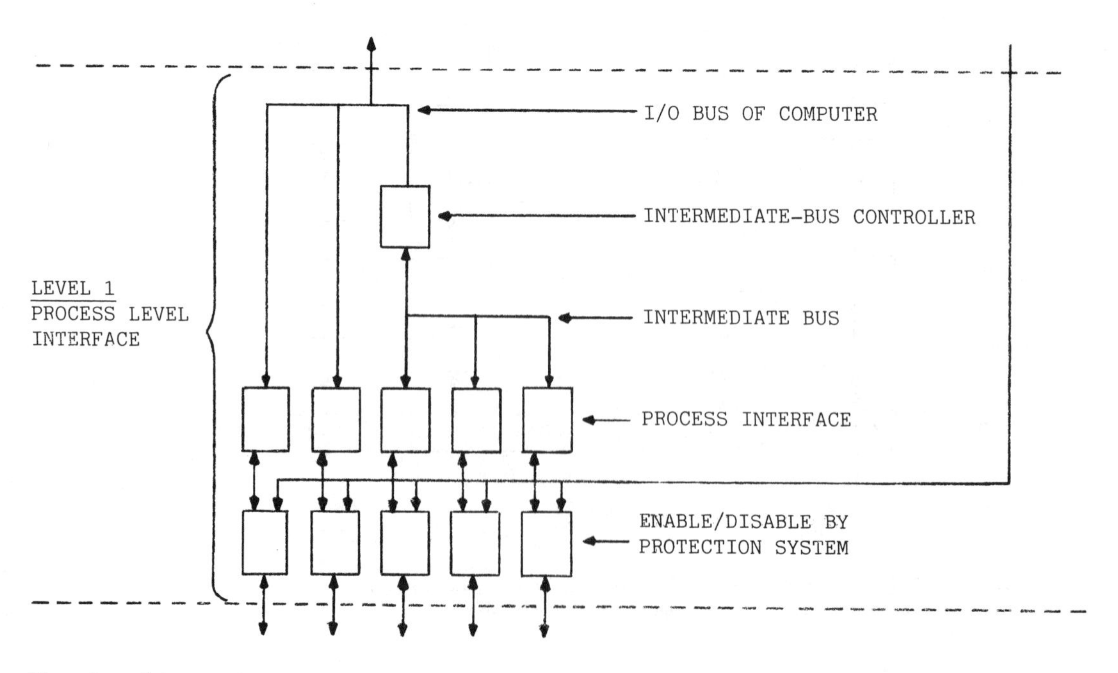

Fig. 5. Sub-Levels of Bit Level in Process Leg, Showing Intermediate Bus

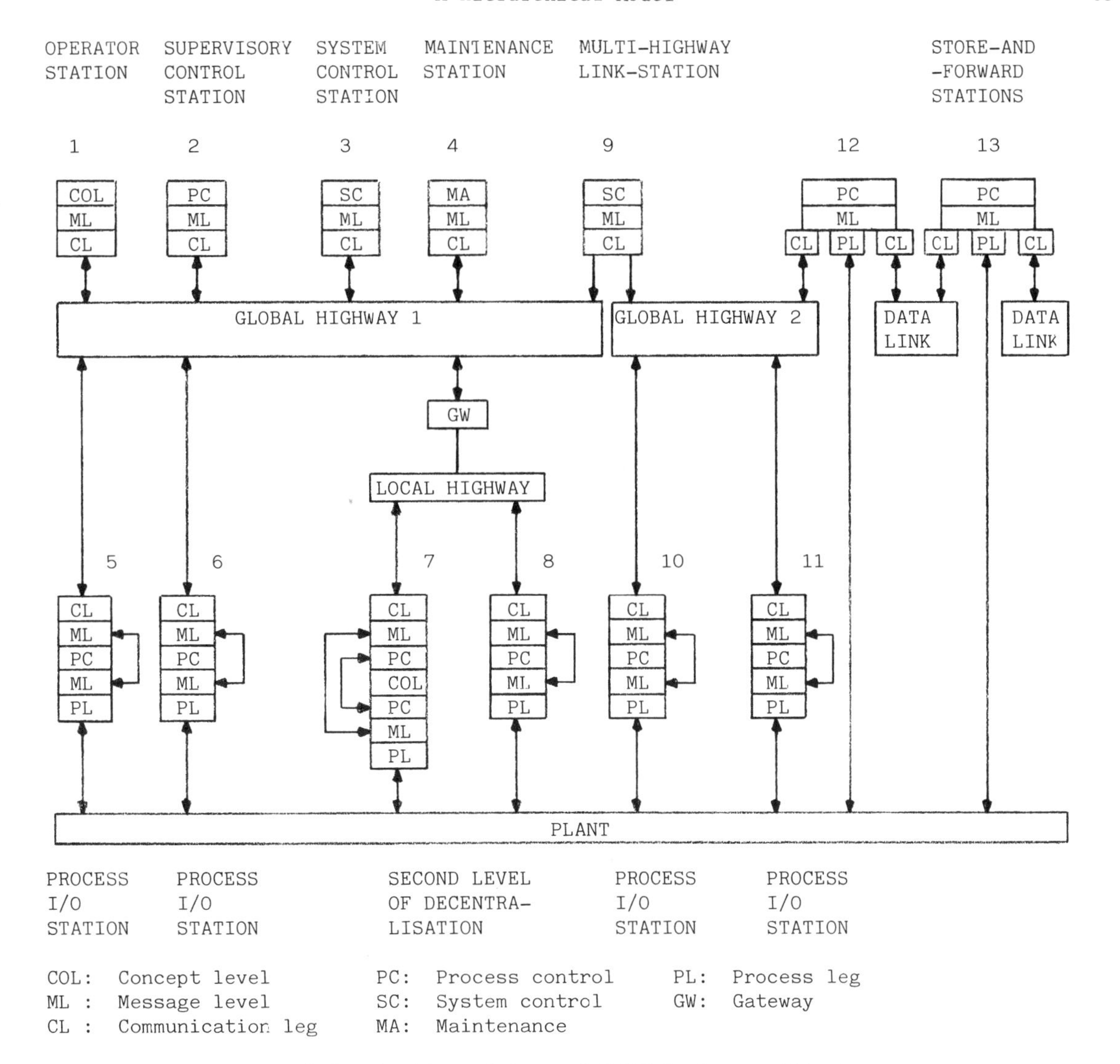

Fig. 6. A Hypothetical Combination of Several MCPC System-architectures

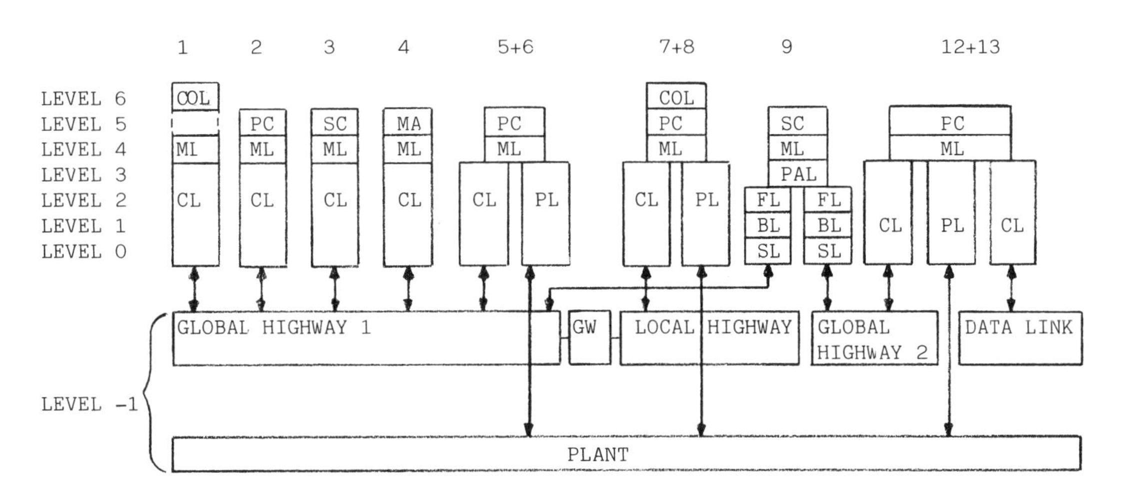

COL: Concept level
ML : Message level
CL : Communications leg
PC: Process control
SC: System control
MA: Maintenance
PL : Process leg
PAL: Packet level
BL : Bit level
SL: Signal level
GW: Gateway
FL: Frame level

Fig. 7. Stratification of the System of Fig. 6

DISCUSSION

Gellie: I would like to refer back to the question of the OSI model. In the early days when we were trying to develop a structure for communications we wrestled long and hard with the problem of logical versus functional and physical versus logical distribution of levels in the hierarchy and I think probably more time was spent on that than on any other area. But I would have thought that OSI got it fairly correct. You made the point, in passing, that you did not think that the OSI model was laid out in levels of abstraction, but rather levels of functionality and if that is true, then you are saying that they did not get it right - whereas I thought they had.

Gertenbach: I am not saying that they did not get it right. The idea in my model is that one looks at the computer system as an information processor. It takes information up from the lower levels to the higher level. The idea was to get the implementation out of the model. So therefore the levels were chosen to reflect how data is used, and the OSI model was seen as a functional hierarchy, described in terms of the service which one level provides to the other. If we talk, for example, about the session layer - then where exactly does this fit in? I see the session layer as an abstraction of interconnections, since this layer manages the communication between the various entities. As has been said in the paper, the interconnection abstraction actually is implementation technology, in that it tells you how to interconnect the various models. I see the that the OSI model does fit into the area, but not as levels of abstraction.

Gellie: Where in your diagram would you see link access management?

Gertenbach: This occurs at the frame level.

Harrison: It seems to me that you have started out with the goal of attempting to match the structure of the OSI model. The OSI model is fundamentally aimed at describing a serial process and it has been put together in such a way that it models this very well. The model does not seem to take into account the parallelism, or the real-time aspects, which it should. I am not certain if you can stick to the pure sequential idea, that the OSI sets out to model, because of the inherent parallelism in the real distributed system.

Gertenbach: Parallelism is actually modelled in terms of decision- making units, or functional units. These are organized into decision- making hierarchies, consisting of functional units, and are distributed in three-dimensional space. Our approach was not to match OSI closely, but rather to look at the total situation. It was also aimed at looking at communication with the plant. It was interesting to make an analogy between what happens in terms of the communications with the plant and communication between computers. The objective was to retain OSI concepts, but to get more clarity as to exactly where to put equipment in the total computer system complex.

Sloman: In Dr Gertenbach's model, he mentioned that he did not think that the OSI model dealt with operating systems. I would like to know whether he has any further thoughts on this?

Gertenbach: The problem with operating systems arises when you consider a process interface in an interface rack and which is independently microprocessor-controlled. This might have its own small operating system and this interface is somewhere - possibly at a specific interface level. The question is - where do you model it? The idea in our work is that on every level of implementation one has got some form of operating principle. On the component level we talk about electronics, on the machine organization level we talk about state machines. The model tells you how to distribute work between various processors.

Sloman: Isn't it in fact hierarchical, but orthoganol to the layers, so that you can, in practice, use the various layers of an operating system almost anywhere within a particular layer of the OSI model?

Gertenbach: That is exactly what the levels of implementation are. They are actually orthoganol - they can model every level of abstraction and every decision unit can be defined to the level of implementation.

Kopetz: I would like to know whether you can analyse your model in terms of its behaviour under fault conditions and provide, for example, a method of locating a fault which occurs in a given sub-system. Would it be difficult to diagnose these faults in your conceptual model? How would the model behave under such fault conditions?

Gertenbach: I think one would have a separate hierarchy for normal operations and another for fault conditions. So in the "good" model you would have to have a system which detects the faulty conditions and which can then switch to the alternate model. So the decision making hierarchy can be imposed onto these various layers.

Sabi-Sabi, South Africa, 1983

DISTRIBUTED VS. CENTRALIZED CONTROL FOR PROFIT

M. Maxwell

Control Systems Engineering, Colgate-Palmolive Company, Jersey City, New Jersey, USA

Abstract. Distributed control started out as an expensive replacement for long thick bundles of wire in large systems. Recently, low cost communication components have appeared which make distributed systems more attractive for smaller installations. Experience with computers in real-time direct-digital control, coupled with new economies in computer components as well as communication components force us to rethink the way we use computers in process control.

Keywords. Computer control; digital computer applications; direct digital control; textile industry.

INTRODUCTION AND PERSPECTIVE

The views expressed in this paper reflect experience in Colgate-Palmolive, in applying minicomputers to the control of existing processes and machines; replacing or augmenting manual and automatic control systems. The experience covers 15 years in which our group assembled, programmed and placed in operation more than twenty minicomputer systems worldwide.

All of these computers were funded on predicted profits from savings in raw materials, operating costs and maintenance expense, sufficient to pay back the investment in two years.

Savings are mere excuses to install computers. After the process has operated with the computer for some time, more important benefits become apparent. The computer has changed, radically, the day-to-day manufacture of detergents, introducing new flexibility, speed and product uniformity.

Radical change starts when programmer-engineers learn the process and its problems from the operating people. It continues when these groups engage in a cooperative start-up. It is perpetuated when the system is flexible.

The essence of flexibility is programmability. The essential medium is Language. The hardware exists only to serve the program. The agony of finding appropriate computer systems is only a background activity whose objective is to provide maximum utility for the plant through implementation of the right language.

The selection of Distributed vs. Centralized control is a part of this activity in order of importance roughly as follows:

1. Instrument reliability
2. System language
3. Programming method
4. System size
5. System Configuration (Distributed vs. Centralized)

Although it is at the bottom of the list, configuration is important because it affects design, installation and maintenance, in depth.

There has never been a question of whether we would use distributed systems. The question has always been; "When?". The answer has always been, "When the benefits pay for the extra cost in our applications".

ECONOMICS

I am sure that in industry, the impetus for Distributed Control like everything else is economic. Distributed control was first developed for installations where thousands of instruments covered square kilometers of area, or where central processors were overloaded. It was economic for huge systems because it saved miles of wire. But what about systems where one hundred instruments cover 1,000 square meters and a central mini is capable of executing all the foreseen programs with no signs of distress?

We asked this question early in the design of each retrofit project. The answer was, "too expensive". But each time the factors affecting the choice were found to be changing in favor of distributed, for example:

1. Wiring costs were always increasing.
2. New communications hardware continued to emerge.
3. We were always finding new uses for our in-house system.
4. We wanted to continue to make our system more versatile; more powerful.

CHANGING TIMES

As we continued to examine this question, the problem was taken out of our hands. From retrofitting we were propelled into the design of new manufacturing facilities, of which computers were perceived as a required part, because they are broadly understood to have the greatest known potential for process control. In the case of a new plant, the evaluation of "Distributed vs. Centralized" is entirely different than in a retrofit.

Here is the chance of a lifetime: to design the plant specifically to be controlled by this magnificent tool, the computer, to help promote the aims of the business.

The emphasis is now on using computers to simplify plant design and installation. In this, every electromagnetic logic device can be replaced by computer logic. When logic is done in a high-level language, the plant is poised for reprogramming whenever needed: to improve excellence and uniformity of product, to provide carefully-summarized operating and management information and to provide responsiveness to changes in "conditions" such as raw material supply and market requirements. The new emphasis should be met without sacrificing attention to the savings initially responsible for computer investment. Also, every new area for profitable plant performance should be explored for computer implications.

Profits result from maximized marketing skill (sales volume/expenditure ratio), product performance/price ratio, manufacturing and distribution quality/cost ratios, wise capital investment, speed of implementing the investment.

Computer skills must be focused to help channel capital investment away from mechanical controls into essential equipment to simplify plant design so that construction is faster.

In this, Distributed vs. Centralized is a key decision.

I would now like to explore the decision-making process in two differing applications in the Colgate-Palmolive Company.

APPLICATION I: COTTON BLEACHING

The bleaching process is horizontally distributed; that is, a number of similar, parallel operations are normally in progress in identical units.

Cotton is bleached in specialized pressure vessels called kiers. In one plant, four or five different bleaching cycles may be used depending on type, grade, use and customer. Ideally, all the kiers would be capable of following any one of these cycles.

The market for cotton is competitive. Moreover, a major customer can decide, at any time, to bleach his own. Thus, one needs to be alert for new markets. Profit-oriented thinkers will recognize the competitive opportunity afforded by a control system that can produce a variety of grades of cotton at low cost, maintain accurate, credible, concise records of cycle parameters for customer relations, maintain raw materials summaries for cost accounting, and be poised to react quickly to market changes!

Whether or not to control bleaching by computer is no longer a question. Economics and technology make the computer a must. Also, we can dismiss the question of language by stating that "BASIC" fulfills, best of all languages, the need to react quickly to changes in requirements. So we can proceed to the point of the paper: Distributed Vs. Centralized.

CENTRALIZED BLEACH CONTROL

A brief look at centralized control will serve to describe the application of the computer.

One sixteen-bit mini with 64K words can control as many as 12 kiers at once, storing recipes for the various bleaching cycles, enabling any kier to follow any recipe at any time.

The same mini can consolidate batch records for all twelve kiers, on one printout.

This arrangement works very well, but it has these disadvantages:

1. The input/output wiring between the computer and the kiers is expensive.
2. If the central computer fails, all the bleach cycles must be finished by hand, or started again from the beginning.
3. A manual backup system for each kier is needed to finish the batch.
4. The crew size for a computerized bleachery will be too small to finish more than one manual batch at a time.

WIRING THE CENTRALIZED CONTROL

For twelve kiers, the distance between computer and kier would average about 30 meters. Each kier would be connected to the computer interface by about sixty wires.

Another sixty wires, 30 meters long are needed for shared equipment such as reagent and utility pumps.

We then have thirteen x sixty wires x thirty meters or 23,400 meters of wire. To protect this wire, a sixty meter conduit run is required, with thirteen major branches. However, inputs must be separated from outputs, to avoid noise pickup. Therefore, two conduit runs are needed. This doubles the length of conduit required.

As a result of these considerations, wiring cost will exceed $100,000.

There is an even more important facet to consider: Checkout of instrument and control wiring. This is the last item on the startup list. How many days does this activity consume when the two ends of hundreds of wires are thirty meters apart? Perhaps a week. Perhaps longer. These are days in which the new installation will make no profit!

FAILURES

In operation, it is inevitable that failures will occur. Although most of these failures will be in the mechanical equipment, there will be an occasional computer failure. Since our policy is to install a spare computer and interface, the resulting downtime may be a matter of minutes. In rare circumstances a failure might result in prolonged downtime.

Since downtime is unpredictable, the usual procedure is to finish as many of the bleaches as possible. However, the crew of a computerized bleachery is too small to finish half the bleaches.

In a prolonged shutdown the unfinished ones must be started from the beginning; wasting chemicals, energy, time, profits!

DISTRIBUTED BLEACH CONTROL

Let us examine distributed control starting with a local microprocessor which can be programmed in Basic, and has its own strip printer and process input/output capability. One can be installed at each kier. This micro can control and record each operation. It can be made to communicate with a central mini in a full-duplex serial mode, so it can receive bleach recipes and send status information.

Now the central mini has the following tasks:

- Store all recipes and download to each micro as required.
- Poll all micros for bleach status information.
- Generate consolidated bleach reports for all kiers.
- Keep records of reagents and utilities consumed.
- Arbitrate simultaneous demands for reagents or utilities.

The physical differences in the process connections are as follows:

- The main wiring link between the computer and the kiers is reduced from hundreds of wires to two twisted pairs.
- The mini interface is replaced by communications equipment.
- The traditional manual backup is replaced by a micro at each kier.
- The control wires are short; easy to check.

The operational differences on central mini failure are:

- All bleaches proceed according to program instead of manually or not at all.
- Bleach records are kept individually until the central mini is restored.
- Automatic arbitration ceases, so that in some cases too many kiers might demand bleach or steam at once. This would slow down the cycles, but, with manual arbitration, efficient automatic operation could continue without the central computer.

COST COMPARISON

We should compare costs of Distributed vs. Centralized in this case.

	Centralized	Distributed
Mini	$ 6,000	$ 6,000
Interface	10,000	-
Printer, Cabinet, Misc.	3,000	3,000
Communications	-	4,000
Main Wiring Runs	50,000	10,000
Local wiring (12 units)	50,000	30,000
Microprocessors (12 units, interfaced)	-	48,000
Engineering	10,000	10,000
Applications Program	30,000	30,000
Software (one time for new design)	-	10,000
Spares	12,000	9,000
	$171,000	$160,000

I would hesitate to draw the conclusion that Distributed costs less, because this estimate was completed in less than thirty minutes, from memory of component prices. Every price is inaccurate, but the conclusion is absolute: Distributed and Centralized control will cost nearly the same for this size project. The advantages of distributed control are free.

APPLICATION II: TOILET SOAP MAKING

The second case is a control system for a new continuous toilet-soap making facility. Its operation is vertically distributed; that is, it proceeds in sequential steps.

First, fats and oils are bleached and deodorized. Then they are treated with a reagent containing salt and caustic soda. The result is washed, settled and centrifuged to remove the spent reagent. In the spent reagent, the remaining salt and the glycerine produced by the saponification process are separated and concentrated by evaporation.

The soap is extruded as noodles and is now ready for mechanical processing into cakes. In this, it is mixed with bateriostat, color and perfume, ground, and extruded again; this time as a bar. Next it is cut into cakes and pressed into shapes designed to please the ladies. Finally, the cakes are dressed in attractive wrappers, packed in cartons and palletized for distribution.

The soap making chemistry is fixed for many products, except for improvements, which are always possible. However, the mixture of fats and oils may be changed in response to supply factors, or marketing strategy.

In mechanical processing, a variety of shapes, colors and scents may be used in response to marketing strategies, market trends or competition.

The conventional continuous soap plant design is founded on considerable process knowledge but is supplied with an antiquated control system consisting of relays, pneumatic control loops, etc. We are about to build a new one in which we plan to substitute a computer for all these devices. The necessary functions will occupy perhaps 20 percent of the available memory and execution time, leaving the rest for purposes we have only begun to imagine. Having made the decision to computerize we investigate the questions with which this paper deals, with the following results.

CENTRALIZED PROCESSING - DISTRIBUTED INPUT/OUTPUT

With a lightly-loaded computer there is no need to distribute processing capability. But, since, as we saw in our previous example, distributed costs have become similar to Centralized, we need to compare the alternatives functionally.

Our principal interest in a distributed system is its wiring implications. We perceive the following advantages.

SIMPLIFIED WIRING DESIGN AND INSTALLATION

a. All instrument and control wiring runs are local. Long fat bundles of wires do not exist. Wiring design time is shortened. Wiring installation time is shortened. Wiring cost is decreased.

b. It is possible to interlock all motor starters via software; to start and stop the plant in software, eliminating almost all pushbuttons. All starter control circuits can then be made ultra-simple and identical. Relay logic is replaced by "Basic" program statements; a direct translation of the start-stop control of the plant. The remaining pushbuttons except those for emergency and maintenance can be wired to the nearest input/output port and connected to their starters via software. The result is further reduction in wiring cost and time. (Emergency and maintenance pushbuttons are hard-wired to the starters.)

c. Without controllers, relay logic or pushbuttons, instrument and control panels are replaced, not augmented, by a keyboard printer and a CRT.

d. As a result, elapsed time for control system design and panel fabrication and assembly disappear from the construction schedule. Control-system design proceeds in parallel with other design phases.

SIMPLIFIED MAINTENANCE

a. We will have replaced pushbuttons, relays, pneumatic transmitters and controllers by electronic transmitters, A/D, D/A & I/P converters, communications loops and a computer.

b. Internally, the new system is vastly more complicated, but easier to maintain because it is modular. It presents trivial problems to the electrician, who simply replaces modules.

c. Indicating lights and diagnostic programs locate the defective module.

d. There is no need to analyze faults in logic because the logic is in the program which is either working or not. A built-in diagnostic and a spare computer reduce all control maintenance to one operation. A faulty program (e.g. one with a dropped bit) is replaced in two minutes from a cassette.

e. All motor control circuits are simple and identical, operable by keyboard diagnostic code. The electrician need not know the interlock sequence. There are no wiring connections hidden away in remote boxes.

f. The expected results are prolonged time between failures; and quicker recovery. This kind of performance saves thousands of dollars in annual operating cost to increase the margin between cost and income. This is profit!

SUMMARY

Through economic justification, we have made computers a part of our manufacturing technology. The knowledge thus gained, and the economic forces which make computers irresistible, have allowed us to see new-plant design in a totally non-traditional way.

We see mechanical simplicity achievable by transferring complexity from electrical-pneumatic control systems to a small box.

Distributed control contributes to mechanical simplicity by transferring complexity from networks of wires to small, replaceable modules.

The result is a plant which can be designed faster, built faster, changed faster, fixed faster: A plant whose performance might be improved dramatically, at any time, by ingenuity of mind, without necessarily making a single physical change.

This plant is geared to making profit.

DISCUSSION

Harrison: I would like to direct a general question at everybody relating to Mr Maxwell's paper in which he brought up the interesting point "Why do you choose distributed systems?" His choice was on the basis that it is not any more expensive, so we might as well take that route. I would like to know if other attendees have run up against this problem, or if they have actually done comparisons.

Sommer: I think if you want practical experience, our practical data can give you the answer as follows: I feel that the problem of down-time and of evaluating how much you lose in production because you did not have the computer available, then the answer to the question becomes apparent. I feel that if we have all data distributed then only a partial failure will occur, and this partial failure will only cause a reduction in production. To put figures to this, in other words comparing a distributed system to a centralized one, our experience has shown that going from one central computer to three smaller machines results in an improvement in performance in the region of 20%.

Maxwell: Our thought is that in some processes it is necessary to have, in fact it is desirable to have, distributed control for a variety of reasons and I think that I considered some of these in my paper. But I am also certain that in some other processes it is not necessary! I can think of instances when we were running three simultaneous processes on one central machine with no distribution at all. In this structure we could lose a machine and could be back on line in a period of time in which no production was lost. Clearly the question relates to the process being controlled. If we have sufficient storage tanks which we can run on for perhaps one or two hours before having problems caused by a computer failure, then the centralized versus distributed problem is put into a different perspective. We have found that we can probably be back on line in 10 to 15 minutes in a case of a failure. In practice we keep a single spare, operating, CPU on the shelf - or if we are far away from the repair depot, maybe two spare CPUs. This means that we can quickly get our system back on line. In many of our applications a broken field wire is really more important to us, and more critical to us, than a failed CPU.

Heher: I would like to comment on AECI's experience in the chemical industry. We have found in analysing a number of plants that we are building that the cross-over point to distributed systems occurs at about R20 000. Below that discrete analog systems appear to be cheaper and above that distributed systems were found to have an advantage depending, of course, on the type of plant being controlled.

Byrne: I don't feel it is reasonable to do any kind of simple comparison in terms of centralized versus distributed. There are many cases where centralized control will not be cheap at any price in terms of the ramification to the process plant on failure of that device. I am inclined to feel that any general type of comment on the subject is not particularly relevant.

LARGE SCALE CONTROL SYSTEM FOR THE MOST ADVANCED HOT STRIP MILL

Makoto Mihara*, Akinobu Ogasawara*, Chikao Imamichi and Atsushi Inamoto****

**Plant Engineering Office, Equipment Department, Nippon Steel Corp., Edamitsu, Yawata Higashi-ku, Kitakyushu-shi, Japan*
***Computer Systems Works, Mitsubishi Electric Corp., 325 Kamimachiya, Kamakura-shi, Japan*

ABSTRACT

This paper will discuss the large scale control system and concentrate attention on the system structure and system design philosophy, with reference to the overall control system of Nippon Steel Yawata Hot Strip Mill. This is one of the most recent and advanced hot strip mill installations, of which requirements on the control system were 1) Large scale process data handling 2) Plant operation based on complex mathematical model 3) Revolutional expansion of digital control and so on.

To respond to these requirements, the physical system structure is composed of four SCC computers, five DDC computers, nine FEPs (Front-End Processor) and thirty PCs (Programmable Controller). The four loops of data-way are also involved for the data communication among the processors. The specific features of this control system, are that the total system is designed to be divided into three levels, say SCC level (with FEPs), DDC 1 level (DDC by computer) and DDC 2 level (DDC by PCs), and that one of the data-way loops, SCC data-way, is assigned for process-information data-base and the other data-way, DDC data-way, for process control data-base.

The most fundamental design philosophy, the main theme of this article, is "Load distribution and resource sharing", and the following items will be discussed in detail:

1) Three level hierarchy control system philosophy.
2) Coupling and optimal load sharing among SCC computers and among DDC computers as well.
3) Load distribution of SCC computers to FEPs.
4) Sharing of process input/output gears and/or other devices and equipment, among computers and PCs.

Keywords. Hot strip mill; large-scale system; computer application; computer evaluation; steel industry; distributed computer system; distributed control system.

INTRODUCTION

In recent years, scale of control system is becoming very large, responding to the requirements of total computerization and large-scale information network accompanied with this computerization.
We intend to describe design approaches to the large scale control system based on our experience of the hot strip mill plant.

History of hot strip mill control systems is relatively old in the control computer fields. Until 1970, the typical control systems were constituted by one central computer and wired logics. When mini-computers appeared, some of wired logics (e.g. automatic position controllers, automatic gauge controllers, etc.) had been replaced with mini-computers. Thus, direct digital control systems in iron and steel application fields made remarkable advances in the period from 1970 to 1975. Programmable controllers (abbr. PC) appeared in the same period. They drove away relay panels at first, then mini-computers the next, after gaining arithmetic operation ability.
Typical control systems in this period are star-configurated with one central computer and mini-computers or programmable controllers. Central computer and lower level controllers (mini-computer or PC) are linked with process input/output or channels.
From 1975 to 1980, distributed control systems were developed with computers and PC's. Dataway was introduced to connect distributed computers and PC's. These systems, considered as two level hierarchy systems, have been very widely applied in the variety of application fields. And the most recent tendencies in these fields, have been proved to be large-scale system approaches.

In early spring of 1980, Nippon Steel Corporation began the construction of a new hot strip mill plant. Not only the most advanced technology of rolling, but also the newest technology of control had been requested for the new plant.
As a result, twenty computers and thirty PC's are installed together with three loops of optical fiber data-way and one loop of coaxial cable data-way. In spring of 1982, the plant was put into operation successfully. This paper describes about the total control system of this plant. Outline of the hot strip mill plant is described in the next section. Then the philosophy of system design and detailed discussions on what is the optimal number of levels in this hierarchy system, how to allocate control functions optimally between computers, and how to decide CPU load distribution, are presented.
In the succeeding sections, the control system configuration, optimally designed according to the above discussions is explained in detail and finally, the evaluation studies based on actual data are presented.

THE HOT STRIP MILL PLANT

The hot strip mill plant

The plant where we introduced the total control system is located in Yawata Works of Nippon Steel Corporation.
This new plant is developed to be the model plant of 21st century in this field, to which Nippon Steel has applied the most advanced facilities and original technology. It is highly automated by large scale computer systems and a number of programmable controllers.

The basic concepts of total control systems are;

1) High quality control and stable plant operation with widely computerized automation.
2) Schedule free rolling by computerized information network and direct coupling of the hot strip mill plant with the steel making plant.
3) Sophisticated rolling with the newest technology line as 6-high mill.
4) Complete energy saving with especially designed equipment and control techniques.

The hot strip mill plant is located in between continuous casting plant and hot coil finishing plant. They are combined with roller tables or chain conveyers. TABLE 1 shows the summary of the plant.
We call the hot strip mill plant with abbreviation YNH (Yawata New Hot Strip Mill).

The computer hierarchy system

YNH control computer system is fabricated in the computer hierarchy system for total plant operation. Fig. 2 shows the position of YNH control computer system in the hierarchy. Production control computers, which are in higher level than hot strip mill control computers, form three level hierarchy as shown in Fig. 1.

In this hierarchy system, YNH control computer system must communicate with the other computers, a great amount of information frequently as shown in TABLE 2.
Because these computers control all of material flow in the plants, all necessary information for production control and technical data are transferred through the computer network within the expected time.

Features of YNH plant operation

1) One man operation aided by fully automated control system and CRT display. (In the conventional hot strip mill plant, there are four to six operators engaged in each pulpit to operate the plant and monitor the instrumentation.)
2) No hard copy operation. Operators have no hard copied rolling instructions nor data sheets. They look at CRT display to know rolling instructions or push CRT key-board to enter

data.

3) Accurate size, shape and metallurgical quality control with pure mathematical model. Simplified model equations conventionally used had a defect that its applicable span is rather narrow. To improve this defect, pure mathematical model equations are applied.

TABLE 1 Summary of Hot Strip Mill Plant

1.	Productivity	380,000 ton/month
2.	Slab Size	250mm × 11.3m Max. 40 ton
3.	Coil Size	(1.2mm - 25mm) × (550mm - 1,550mm)
4.	Steel Grade	Carbon Steel, Silicon Steel, Stainless Steel, Special Steel
5.	Features of Operation	Hot slab charging to the reheating furnace. Schedule free rolling. Coil division with crop shear.
6.	Features of facilities	Roughing mill with great width reduction edger. Reheating furnace especially designed for energy saving. Six-high mill with oil-pressured positionner.

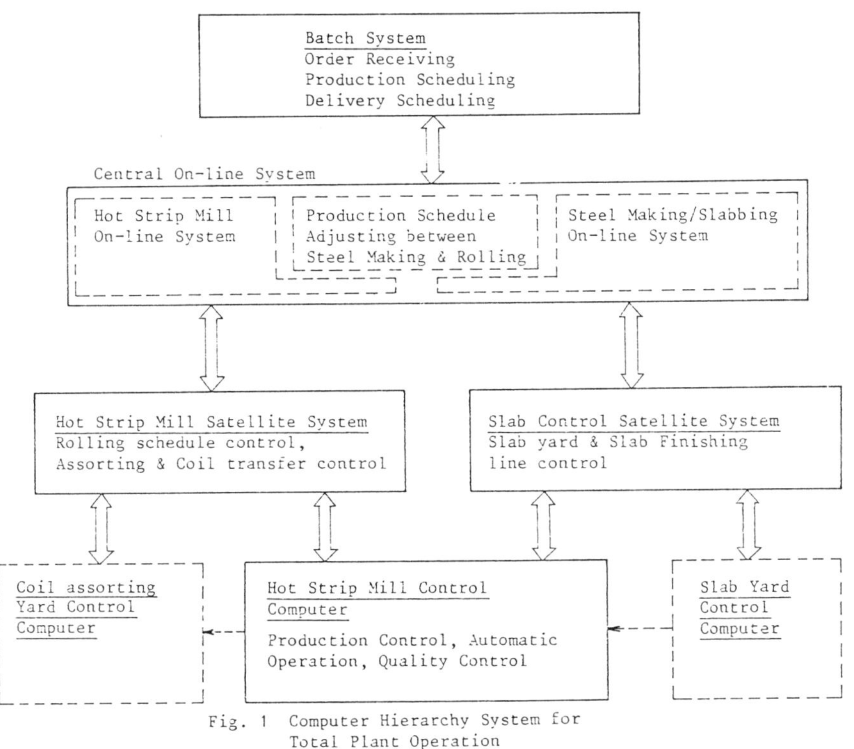

Fig. 1 Computer Hierarchy System for Total Plant Operation

TABLE 2 Data Transmission Between YNH System and the other Computer System

Transmitting Direction	Data Items	Message Length	Transaction Frequency
S.M.S. On-line to YNH	Slab delivery schedule & slab data	50 words	Max. 3,000/day
YNH to S.M.S. On-line	Slab delivery request	25 words	Max. 200/day
H.S.M. On-line to YNH	Rolling instruction	14,000 words	Max. 30/day
YNH to H.S.M. On-line	Production report	100 words	Max. 5,000/day
Between YNH & F'CE instru-mentation	Combustion control data and references	500 words	Max. 1,400/day
Between YNH & CDT	Command & data for facility diagnosis	50 words	Max. 100,000/day

(abbr. S.M.S. ; Steel Making/Slabbing, H.S.M. ; Hot Strip Mill)

Requirements for the computer system response.

1) Conventional display panels should be replaced with CRT display and part of indicators and/or recorders also replaced with CRT display. So that operating gears and panels should be reduced in numbers and size as small as possible to make it easier to reach them by one operator.
CRT display response time should be 1/2 second for first response and 1 to 2 seconds for completed display to keep responses within the thresholds of human expectation.
2) Schedule free operation requires timely slab supply to the hot strip mill plant. For this purpose, YNH computer system must communicate with the steel making plant production control computer and the hot strip mill production control computer within proper response time.

Pure mathematical model based control.

To realize accurate size, shape and metallurgical quality, the following mathematical equation based control must be implemented by computer, in addition to the other control functions.

1) Combustion control of reheating furnace estimating slab temperature with partial differential equations.
2) Rougher and finisher mill set up with theoretical model equations and load balance optimization with dynamic programming algorithm
3) Strip temperature control with theoretical cooling equations.
4) Diagnosis of rotating machines with dynamic monitoring and vibration wave form analysis.
5) Diagnosis of oil servo mechanism with response characteristics analysis.
6) Control results data gathering and its analysis.

Each program size, execution frequency and required response time are shown in Table 3.

TABLE 3 Control Items & Models of SCC

Items	Model Size	Response Time	Frequency
Combustion control	80kW	Ave. 5sec.	Max. 4/min.
Rougher set up	100kW	Ave. 3sec.	Max. 1.5/min.
Finisher set up	140kW	Ave. 3sec.	Max. 2/min.
Finishing temperature set up	15kW	Ave. 1sec.	Max. 1/min.
Coiler temperature set up	10kW	Ave. 1sec.	Max. 1/min.
Diagnosis	5kW×40	Ave. 2sec.	Max. 10/min.
Model simulation & analysis	100kW	1 - 30sec.	Max. 1/min.

Requirements for the direct digital control.

Control algorithms used in the conventional DDC is mostly simple because of the limitation of calculation time or response time. We have intended to introduce much higher level algorithms such as filter theory, multivariable control theory and gradient method as shown in below.

1) Adding to the ordinary roll force feedback/feedforward AGC and X-ray monitor, estimation of mill spring and coefficient of plasticity with optimum linear filter theory has been introduced as on-line function.
2) Theoretical plasticity model is applied to calculate draft value at each longitudinal control point of rough bars for automatic width control.
3) Strip temperature is controlled with induction edge heater, stand sprays, mill speed and run-out table sprays. Feedback and feedforward control is applied on the basis of temperature sensor signals.
4) Feedback control is applied for steering control on the basis of edge position detector signal and force difference between work side and drive side of mills.
5) Estimating strip shape by characteristic thickness gauge pattern to optimize the shape with gradient method.
6) Estimating strip crown by crown meter signal to optimize the crown by gradient method.
7) Non-intervention control between steering control and thickness control.

Program size, required response time and sampling time for each above mentioned function are summarized in TABLE 4.

TABLE 4 Control Items & Models of DDC

Items	Model Size	Response Time	Frequency
Automatic gauge control	30kW	10 - 15ms	1/20ms
Automatic width control	20kW	20 - 50ms	1/40 - 100ms
Strip temperature control	70kw	50 - 200ms	1/100 - 500ms
Steering control	5kW	5 - 10ms	1/10 - 20ms
Shape control	20kW	0.2 - 1sec.	1/0.5 - 2sec.
Crown control	20kW	0.2 - 1sec.	1/0.5 - 2sec.
Decoupling control	5kW	5 - 10ms	1/20ms

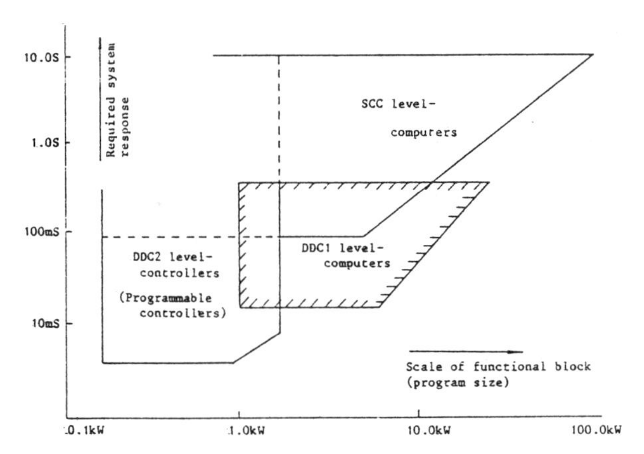

Fig. 2 Evaluation of computers and controllers

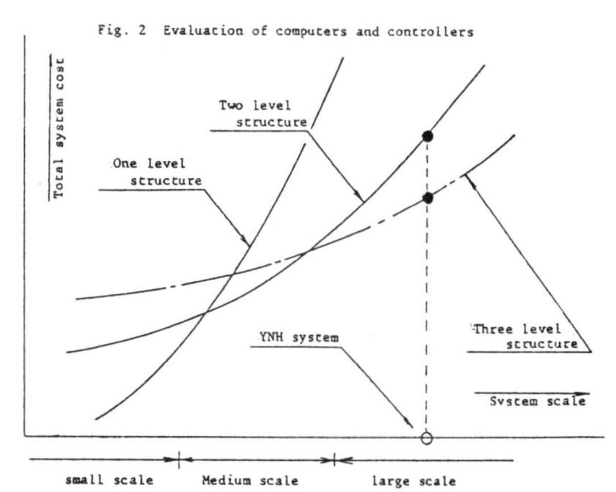

Fig. 3 Total system cost estimation

SYSTEM DESIGN

In order to meet the above-mentioned system requirements and at the same time to challenge the large-scale control system design, new system design concepts should have been established.
The major four concepts are important, which are:

a) To determine the number of levels of total control system hierarchy;
b) To optimally allocate the functional blocks among the computers, which build-up multi-computer system architecture;
c) To distribute the jobs of computers to FEP (front-end-processors);
d) To make the process input/output resources to be shared among computers and programmable controllers (abbr. PC).

Three Level Control System

Generally speaking, not limited to the control system of a hot strip mill plant, the control systems may have hierarchy structure. Up to now, two-level hierarchy control systems seem to be very common. These two levels are:

a) SCC (Supervisory Control Computer) level;
b) DDC (Direct Digital Control) level.

And as for SCC level computers, medium scale SCC computers (process control computers or industrial system computers) are to be applied, while as for DDC level controllers, micro computers (in this paper, programmable controllers) are to be applied.

Now, we are going to review the differences in performance between SCC computers and programmable controllers. For this purpose, we will show an evaluation chart. In Fig. 2, X-axis represents program size of an individual functional block, while Y-axis represents required response time or sampling period. And the enclosed area in the figure shown the application range, of each level.
Investigating the application ranges of SCC computers and plant controllers, the following points may be pointed-out:

a) SCC computers and programmable controllers cover, to some extent, all the required application ranges of total control system hierarchy.
But relatively quick (30 ms∿1 sec) small functions, that range 0.5K to 2K words, are not covered enough.

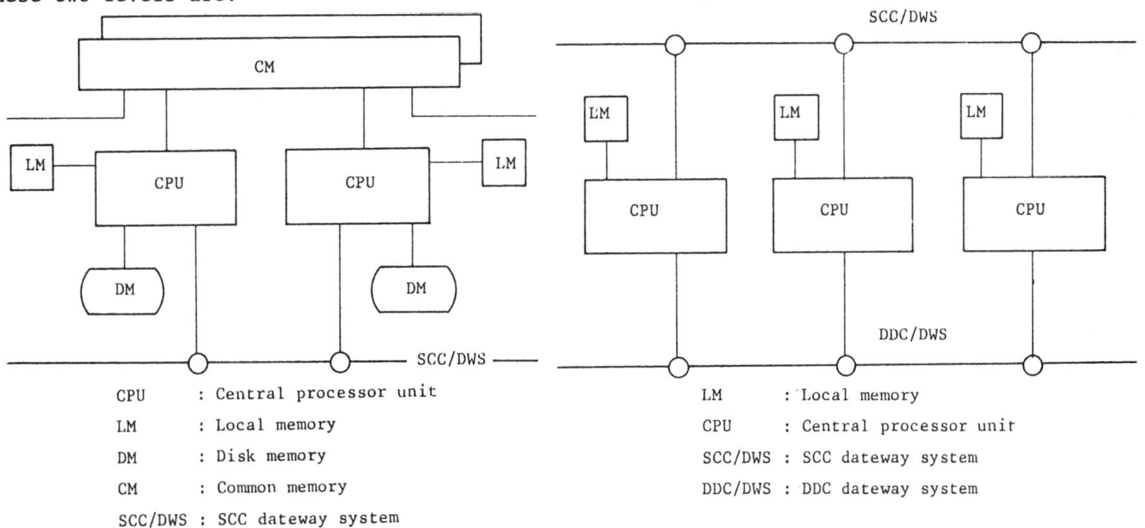

Fig. 4(a) Multi-computer system architecture in SCC-level Fig. 4(b) Multi-computer system architecture in DDC1 level

b) It seems better if we have one more level between SCC computer level and programmable controllers level, considering the new requirements on advanced DDC,as mentioned in the the preceeding section.

When we design the total control system hierarchy, one level control system may be applicable in case of small scale control system. For medium scale control system, two level control system may be suitable, and two level control system may be extended to large-scale control systems.
However, on designing YNH total control system, we cannot stick to the conventional two-level control system, because the system scale is thought to be the largest ever designed, and at the same time, many new requirements are to be added, specially on the mathematical model based control algorithm.

We made cost analysis to determine the hierarchy configuration of YNH system. As the result, we got the conclusion that the three-level control system with DDC1 level for new requirements must be more suitable. In Fig. 3, three curves which represent total system cost vs. system scale are shown by indexing number of levels. Reviewing these two figures, it may be concluded that three level control system may be optimal in case of YNH total control system, considering both system scale and new requirements.

Optimal Functional Allocation

According to the analysis, presented in the previous sub-section, three level control system structure may be chosen, and then the next problem would be how to allocate the required functional blocks among the computers or controllers within the level. This is a kind of multi-computer system design problem.
There are two major categories of interconnecting processors and memories which can be called multiprocessor system (or tightly coupled system) and multicomputer system (or loosely coupled system). In the former the autonomous processors are connected to share main memory units through a distinct interconnection network, while in the latter the processors, with attached local memory units, are connected only by message passing buses to other similar processor-memory pairs. (Arden, 1982)

We selected multi-computer system in this case based on the following several reasons;

(1) Control functions are relatively independent of the others.
(2) Material processing data and process signals may be the information mostly used in common between computers.
(3) Thus, interactions between functions are not so frequent and require relatively slow response time.

Looking at the YNH hardware system structure, the multi-computer system would be featured as Fig. 4 (a) and Fig. 4 (b), for SCC level and DDC1 level, respectively.
As shown in these figures, very identical structure would be prepared for each of CPUs, to constitute the ideal multi-computer system architecture.

The optimal design criteria, applied for SCC level and DDC1 level, were as follows:

(1) The basic design concept concerning multi-computer system should be based on loosely coupled architecture.
(2) The final design criterion is to minimize the CPU load shared for inter-computer data communication, so that total system performance may not be reduced.
(3) Each of computer systems should be well-balanced with respect to a) average CPU load, b) required local memory size, c) required disk memory (especially fixed head disk memory).

In order to formulate the above-mentioned optimal design criteria, we have to introduce a model on the functional structure. The model illustrated in Fig. 5 would be applicable in this case. Here, the total control function is assumed to be divided into a number of functional blocks. And each functional block is assumed to require so much amount of CPU load, local memory size and disk memory. The data communication between functional blocks may exist.

With reference to this model of functional structure, the over-all optimal design problem could be formulated, as in the following equations: (see Kasahara, 1982)
In order to obtain the final solution to this optimal task allocation problem, we have to at first prepare the data tables. These data tables, of which samples are shown in TABLE 5(a) and TABLE 5(b), give us the very fundamental data.

In order to solve mathematically this optimal problem, we have to introduce some sort of non-linear integer programming techniques. However, here, we have obtained the suitable solutions by trial and error method starting from the conventional functional allocation.

(1) The optimal problem is to determine a set of assignment variables, which minimizes the total sum of inter-computer data communication:

$$Q = \sum_{i=1}^{m} \sum_{j=1}^{m} \sum_{k=1}^{n} \sum_{\ell=1}^{n} [Q_{k\ell} * X_{ik} * X_{j\ell}] \qquad (1)$$

(2) Restrictions on inter-computer balance may be expressed as:

$$\left| \sum_{k=1}^{n} \left(M_k * X_{i,k} \right) - \bar{M} \right| \le \Delta M, \quad i = 1, \cdots, m \qquad (2)$$

$$\left| \sum_{k=1}^{n} \left(D_k * X_{i,k} \right) - \bar{D} \right| \le \Delta D, \quad i = 1, \cdots, m \qquad (3)$$

$$\left| \sum_{k=1}^{n} \left(L_k * X_{i,k} \right) - \bar{L} \right| \le \Delta L. \quad i = 1, \cdots, m \qquad (4)$$

where

$$\bar{M} = \frac{1}{m} \sum_{k=1}^{n} M_k, \quad \bar{D} = \frac{1}{m} \sum_{k=1}^{n} D_k, \quad \bar{L} = \frac{1}{m} \sum_{k=1}^{n} L_k. \qquad (5)-(7)$$

(3) Variables are defined as:

X_{ik} (assignment variable)
= 1 : functional block k is assigned on CPU-i.
= 0 : otherwise (8)

$Q_{k\ell}$: amount of data communication from functional block-k to functional block-ℓ

M_k : required local memory size associated with functional block-k.

D_k : required disk memory size associated with functional block-k.

L_k : occupied CPU load associated with functional block-k.

TABLE 5 Data Table for Functional Allocation

	Functional Block	CPU Load	Local Memory Size	Disk Memory Size
T_1	Furnace entry side tracking	L_1	M_1	D_1
T_2	Furnace tracking	L_2	M_2	D_2
T_3	Mill line set up	L_3	M_3	D_3
T_4	Mill line tracking	L_4	M_4	D_4

(a) Data table for each functional block

From \ To		T_1	T_2	T_3	T_4
T_1	Furnace entry side tracking		Q_{12}	Q_{13}	Q_{14}
T_2	Furnace tracking	Q_{21}		Q_{23}	Q_{24}
T_3	Mill line tracking	Q_{31}	Q_{32}		Q_{34}
T_4	Mill line set up	Q_{41}	Q_{42}	Q_{43}	

(b) Data table for inter functional block data communication

CPU Load Distribution

On designing the system configuration of YNH computer system, the basic objectives of introducing the distributed system architecture with SCC dataway system and FEPs (abbr. front-end-processors) are as follows:

(1) To reduce cabling cost associated with peripheral devices, while the reduction of process input/output cabling is realized by the introduction of DDC dataway system.

(2) To distribute tasks of SCC computers among FEPs, for reduction of the CPU load of SCC computers.

Taking process data handling at the SCC level as an example, it will be shown how the data-processing cost is to be reduced by introducing the distributed system. The analysis below is comparison of two system configuration on process input/output which is built on DDC dataway system. The one is a) direct coupling configuration and the other one is b) load-distributed configuration. According to the former configuration, as shown in Fig. 6 (a), SCC computers have direct coupling with DDC dataway system, so that all of the data handling on process input/output may be done in SCC computers. At the latter configuration, as shown in Fig. 6 (b), PIO/FEP is to be prepared, so that the primary parts of process data handling may be done at PIO/FEP.

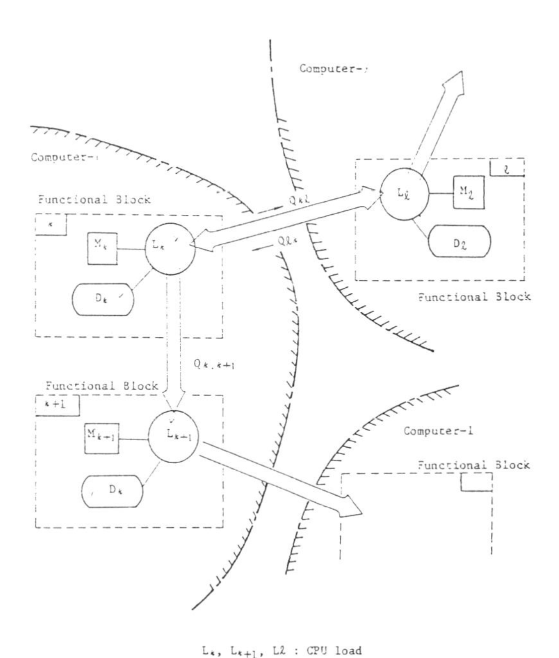

Fig. 5 A model for functional structure

Here, data processing cost is one of the major factors, in choosing either one of these two configurations. And for estimating the data-processing cost of this process data handling, the following premises are required: (see Muto, 1982)

(1) If directly coupled, the data processing cost is proportional to the number of process I/O points;

(2) The system total cost is estimated according to the following equation:

$$Ct = \sum_{i} \alpha i * Pi + \sum_{i,j} Qij$$

Ct : total data processing cost
α_i : cost coefficient at computer-i
P_i : number of process I/O points to be processed at computer-i
Q_{ij}: data communication cost if there exits communication between computer-i and computer-j, Else, Qij=0.

(3) The cost coefficient at SCC computers is much higher than at PIO/FEPs, i.e.

$$\alpha i >> \alpha j \quad i < SCC\ computers \quad j < FEPs$$

The total data processing costs are calculated and graphically shown in Fig. 7, in both direct coupling case and distributed system case. In both cases, number of SCC computers are parameterized.
According to this figure, the distributed system with PIO/FEP is thought to have advantages over the direct coupling in case of YNH system, considering the number of process data to be handled and the number of SCC computers.

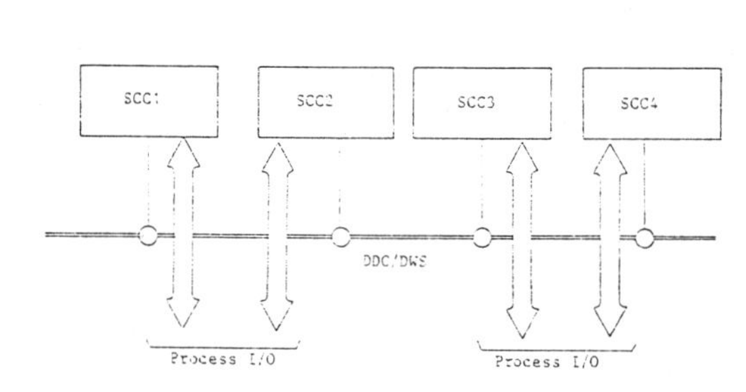

(a) Direct coupling structure

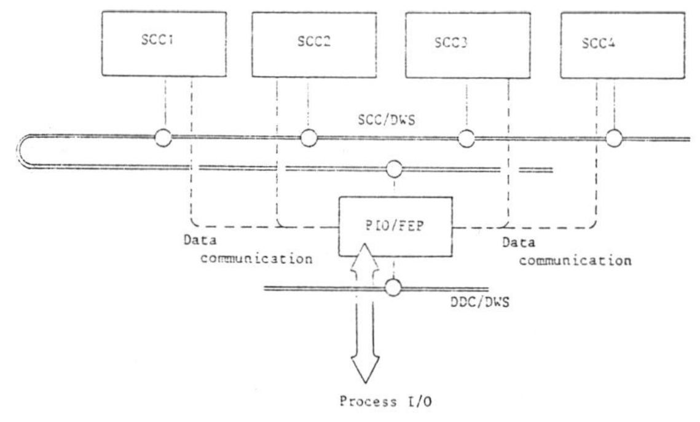

(b) Distributed system structure

Fig. 6 Two system structure on process input/output

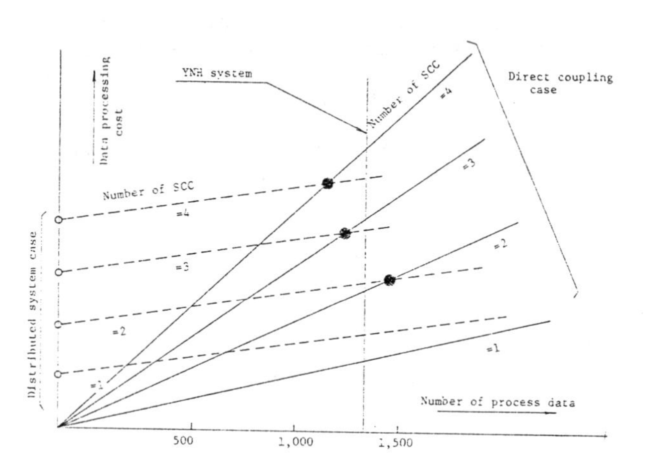

Fig. 7 Data processing cost on process data

SYSTEM CONFIGURATION

The final system configuration designed according to the preceding discussion is described in this section.
There are four supervisory computers (SCC), five direct digital computers (DDC 1 level), six sets of front end processors (included in SCC level) and thirty PC's (DDC 2 level) in YNH control system. SCC and DDC 1 level computers are connected all together with SCC data-way. DDC 1 level computers, front end processsors (FEP), and PC's are connected with DDC data-way. Condition diagnosis technique systems(CDT)and electrical maintenance monitor system (EMM) have each its own data-way. This YNH control system is shown in Fig. 8.
The specification of each model is shown in TABLE 6.

TABLE 6 Hardware Specification

Component	Specification	
SCC & DDC1 computer	model	: MELCOM 350-50/A2500
	fixed add time	: 0.2 μsec.
	memory cycle time	: 0.65 μsec./64 bits
	max. memory size	: 1 M words
	number of instructions	: 250
Front-end processor	model	: MELCOM 350-50/A2500
	fixed add time	: 1.2 μsec.
	memory cycle time	: 0.6 μsec./16 bits
	max. memory size	: 192 K words
	number of instructions	: 164
SCC data way	transmission ratio	: 15.36 M bits/sec.
	communication links	: optical fiber
	max. stations	: 256
DDC data way	transmission ratio	: 15.36 M bits/sec.
	communication links	: optical fiber
	max. stations	: 127

1) The reason why we prepared both SCC data-way and DDC data-way separately is to expect specially high throughput on SCC data-way and good response characteristics on DDC data-way, respectively. YNH system requires two kinds of definitely different characteristics to the communication system. The first one is application-to-application and the second is sensor base application. In the first application, there is a strong requirement for high level functions, such as automatic recovery from errors. On the other hand, in sensor base application, most emphasized requirement is response time.
On the above reason, we chose message type communication with protocol for SCC data-way, and cyclic transmission type communication for DDC data-way.

2) DDC data-way was developed to get enough response time for direct digital control and to offer the access method to common input/output system for computers and PC's.
Stations are furnished with writable memory bank, which is divided into input section and output section. All of the remote inputs image are transferred periodically to each station's input memory section.
Each station's output memory section image is also transferred to remote output stations periodically. When the computer changes the output image of its station compared with preceding cycle, output relays or transistors corresponding to the changed memory bits of the remote station will be energized.
Thus, computers and PC's can use remote input/output system in common without protocol.

3) The framework of functional ranking between three hierarchy levels is determined as shown in TABLE 7, considering performance of computers and PC's. There are three levels, SCC, DDC1, and DDC2. Mini computers (16 bits/word) are used in SCC and DDC1 level.
PC's are used in DDC2 level.
Indexes that characterize each level are response time, function size and control span.
For the convenience of comparison, the above indexes are substituted by OS, control period, tracking scope, algorithm and language in TABLE 7.

4) As the same computer model was chosen for SCC and DDC1 from the view point of performance and maintainability, operating system is different between SCC and DDC1. To get short response time, smaller operating system is applied to DDC1 system.
Control programs of DDC1 reside on main memory for quick response.
DDC1 programs are also written in Industrial FORTRAN, so that development and maintenance of DDC1 control program could be implemented, with equal degree of convenience as SCC level.
DDC1 computers are directly connected to DDC data-way to shorten transmission time delay.

TABLE 7 Functional Ranking of Three Level Hierarchy

Level	Operating System	Control Period	Tracking Scope	Control Algorithm
SCC Level	Disk operating system	Depends on event signal	Zone tracking (master control of DDC1 & DDC2 level)	Mathematical model control
DDC1 Level	Main memory operating system	20 to 500ms sampling period	Local tracking (corresponding to sensor signal)	Mathematical model & feedback/feedforward control
DDC2 Level	No operating system	5 to 100ms sampling period	Local tracking (corresponding to sensor signal)	Feedback & sequence control

5) Functional allocation for SCC and DDC1 computers are shown in TABLE 8.
SCC1 controls furnace entry tables and furnaces.
SCC2 controls mill line starting from furnace exit tables to coiler conveyers.
SCC3 diagnoses mechanical equipment and servo mechanism.
SCC4 monitors electric equipment primarily but in case of the other SCC's down, it will do the all control functions substituting the malfunctioned CPU.

TABLE 8(a) Functional Allocation

Computer System	Functions
SCC 1	Supervisory control of reheating furnace & associated facilities, including slab receiving & transfer, combustion control of furnaces, receiving of rolling schedule.
SCC 2	Supervisory control of mill line, including set up calculation of rougher, finishing coiler & coolant, tracking of coils and data gathering.
SCC 3	Supervisory control of Condition Diagnosis Technic system, including schedule control of diagnosis, data gathering, forecasting & deciding of malfunctioned facilities.
SCC 4	Back-up for SCC 1 to 3. System support functions including monitoring of electric facilities, analysis of gathered data, program development, and RAS functions.
SCC Data-way	Forms supervisory control information network for YNH supervisory control system.

TABLE 8(b) Functional Allocation

Computer System	Functions
DDC 11	Automatic width control (AWC), including rough bars, feed-forward control of width fluctuation, step shift control of width.
DDC 12	Automatic gauge control (AGC), steering control (STC), dynamic set up of finisher DSU), and decoupling control of multi-variables (ADC).
DDC 13	Automatic shape control (ASC), automatic crown control (ACR), and edge heater control (EHC).
DDC 14	Coil temperature control (CTC), finisher temperature control (ATC), and quality judging by gathered data.
DDC 15	Back-up for DDC 11 to DDC 14, and AGC dynamic simulation.
DDC Data-way	Carries process input/output data and DDC references at high transmission efficiency.

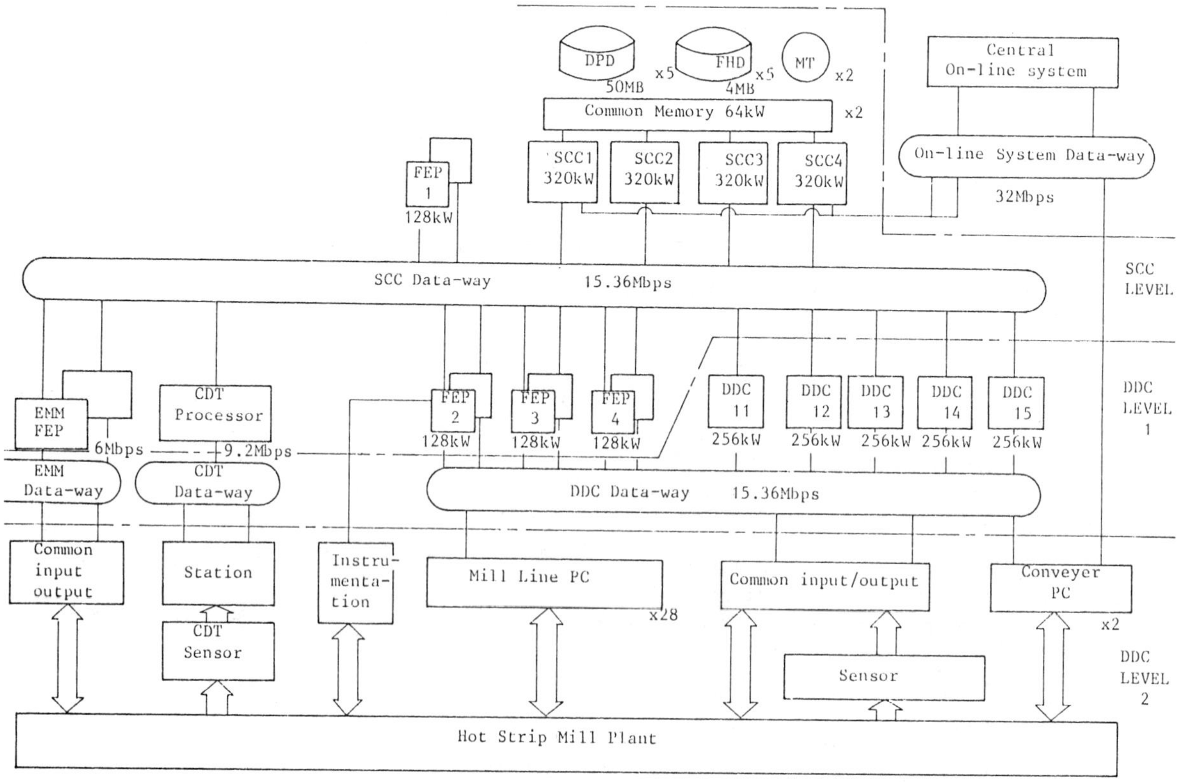

Fig. 8 YNH Total Control System Configuration

EVALUATION

Here, in this section, in order to show how the system design concepts of this YNH total control system have been implemented, some of the evaluation studies are presented, based on the actual field data of YNH system.

CPU load distribution

The field data concerning this matter are shown in TABLE 9(a) and TABLE 9(b) for SCC-level and DDC1 level respectively. These field data show that:

(1) Inter-computer data communications are kept low. Less than 5.0% of total CPU load is shared for inter-computer data communication.

(2) Main memory size, and disk memory size are well balanced. However, to estimate the CPU load for each of the functional blocks may be an easy job. So the deviation of CPU load seems to be higher than designed.

TABLE 9 Evaluation of Functional Allocation

	Average CPU load	Main memory size	Disk memory size	Data Communication To	Data Communication From
SCC1	60%	320	50 MB +4 MB	SCC2 SCC4 FEP	SCC2 SCC4 FEP
SCC2	73%	320	50 MB +4 MB	SCC1 SCC3 DDC1 FEP	SCC1 SCC3 DDC1 FEP
SCC3	55%	320	50 MB +4 MB	SCC2 FEP SCC4	SCC2 FEP SCC4
SCC4	60%	320	100 MB +4 MB	SCC1 SCC3 FEP	SCC1 SCC3 FEP

(a) SCC level

	Average CPU load	Main memory size	Disk memory size	Data Communication To	Data Communication From
DDC11	63%	256	-	SCC2 FEP	SCC2 FEP
DDC12	80%	256	-	SCC2 DDC13 FEP	SCC2 DDC13 FEP
DDC13	40%	256	-	SCC2 DDC12 FEP	SCC2 DDC12 FEP
DDC14	75%	256	-	SCC2 FEP	SCC2 FEP
DDC15	Back-up	256	-		

(b) DDC1 Level

Three level control system

First, in order to show how three level control system actually functions, the strip gauge control system at the finishing mills is to be given as an example. And the functional sharing between the levels and data flows is shown in Fig. 9.

In case of this strip gauge control system, at SCC-computer adaptive control and mill set-up schedule calculation are done basically for every coil. At DDC1-computer, A.G.C. (automatic gauge control) programs run every 20msec. (minimum) and also local tracking is implemented. At DDC2-level, very fast sampling control (sampling time =

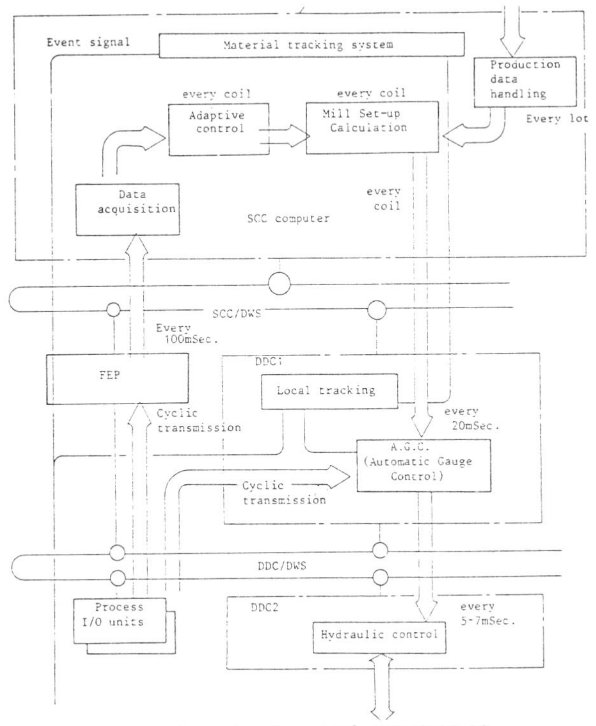

Fig. 9 Three level control structure on strip gauge control

TABLE 10 Evaluation of Three Level Control System

	Required Response Time			Control Program Size			No. of Programs per CPU
	Ave.	Max.	Min.	Ave.	Max.	Min.	
SCC	1 sec	5 sec	200ms	0.5kw	10 kw	0.2kw	200
DDC1	100ms	200ms	20ms	0.5kw	3kw	0.2kw	40
DDC2	40ms	80ms	5ms	0.1kw	0.5kw	0.05kw	

5 ~ 7msec.) is realized. And finally, process data are scanned and preliminary processed at PIO/FEP and those data are transmitted to SCC-computers every 100msec.

Next, in order to justify the three level control system structure on this YNH control system, some statistical data are given in TABLE 10. In this table, required response time and control program size are listed-up for every control level, and number of control programs per CPU are given.
Reviewing this table, it may be concluded that the existence of DDC1 level may well justified.

Process data resource sharing

Let one of the major objectives of introduction of DDC data-way system is to process I/O resources be shared among SCC computers, DDC1 computers and DDC2 controllers. TABLE 11 gives us rough idea on how much this process I/O sharing is implemented.

TABLE 11 Status of Process I/O Sharing

SCC	DDC1	DDC2	Percentage
✓			
	✓		
		✓	
✓	✓		
	✓	✓	
✓		✓	
✓	✓	✓	
		Total	100%

✓ indicates sharing

CONCLUSION

The characteristics of the new hot strip mill automation may be described as follows.
The hot strip mill plant is controlled by the computer hierarchy which includes from production control to process control. Members of the hierarchy system should communicate with others through data link within the expected response time. Either the production control information or operational instructions and processing status information produced or gathered by computers will be fed to each operator by CRT terminals.
Thus the plant can be controlled effectively by one third of operators of the conventional plant.
To achieve more accurate size, better shape and metallurgical quality, pure mathematical models should be solved on-line by the control computers with in 20 ms to 1 sec. response time.
The required control span for the control system is very wide. That includes combustion control of reheating furnace, mill set up, diagnosis of machines and electrical equipment, automatic gauge control, automatic shape control and automatic coil temperature control.
Hierarchy and distribution are the system design philosophy that has been feasible to this control system.
It was reasonable that control computer system was also structured in hierarchical form.
How much hierarchy levels should this control system have, was discussed on the base of cost/performance analysis. We chose three level hierarchy, that is SCC, DDC1 and DCC2.
The next problem was how to distribute or allocate functions for the member CPUs.

Evaluation criteria for this problem were:
1) how to minimize the inter CPU communication load
2) how to balance the CPU load and memory size
To solve this optimal problem, we used trial and error methods instead of mathematical method.
Futher it is shown that process input/output processing on FEP may be preferable to processing them on main computers from the view point of processing cost.
The hardware and software configuration realized according to the preceding design philosophy were explained. Two level data-way is the point of hardware architecture.

Criteria for dividing functions into three levels are shown with indexes that characterize each level.
Then actual function assignment to four SCC computers and five DDC1 computers was described.
Evaluation of YNH system with actual data was developed. Proposition of balancing the CPU load distribution and memory size of each computer may be said satisfied.
Three level hierarchy system is evaluated with response time and program size. An example of functional organization of three level system was also shown. That shows how SCC, DDC1, DDC2 and FEP function in the system.
Operation and maintenance of such a large system is another problem for the designer.
It has passed nearly one year since the YNH system got into operation. So the system design must be reviewed from this point of view as the next step.

REFERENCES

Satyanarayanan, M. (1980). Commercial multiprocessing systems, Computer, 13, 75-95.

Kasahara, H. and Narita, S. (1982) Parallel processing for real time control and simulation of DCCS. IFAC/DCCS-82, Preprint.

Muto, T., Imamichi, C., Inamoto, A. and Kato, S. (1982) Development and quantitative evaluation of distributed sensor base management system. IFAC/DCCS-82. Preprint.

Arden, B. W. and Lee H. (1982) A regular network for multi-computer system. IEEE Trans. Comput., C-31, 60-69.

DISCUSSION

Kopetz: Mr Imamichi discussed the question of balance between memory, communication bandwidth and CPU capacity. Where exactly is this balance and is it possible to obtain numbers for the particular application.

Imamichi: I first discussed the balance of CPU and memory size. There are four SSC computers and, as the paper shows, it may be said that the CPUs are balanced because there are no critical cases in which a CPU is particularly heavily loaded. The memory is balanced since the memory sizes in all four computers are the same. This is particularly important because we have one back-up CPU as a spare and so the hardware configuration should be identical throughout the SSC CPUs. I have not shown the communication balance as there are so many communication components, but in practice we have found that our communication requires less than 5% of CPU time, so we do not have a particular problem in this case.

Harrison: It appears that CPU loading, as shown in the paper, is high - in fact well above 50%. Does this imply that there will be a problem in expanding the system?

Imamichi: In practice we do have CPUs which are operating fairly close to their limit. We need a rapid response time so if the CPU loading became greater, then clearly the response time could become unacceptable. This was expected during the design phase but it was found difficult to separate the various functions being performed.

Heher: Can you give any detail regarding the protocols which were used in the communication system?

Imamichi: The communication data-way is a ring and uses optical fibre transmission. The method which is used for data transfer is that each message is in a fixed position, in packets - in essence the approach is similar to packet insertion.

Harrison: How is the contention between the DDC computers handled?

Imamichi: Input is distributed to any computer at any one time so there is no problem. If these computers wish to send out information to one point at the same time then we have a problem, so we never use the same point for the same output from different CPUs. Output is limited to one CPU.

Harrison: It appears therefore that the system is similar to TDM with the program responsible for assigning slots, in other words keeping track of the slots. The communication system essentially looks like a long shift register.

Kopetz: What is the mechanical interface to the optical fibre and does the fibre in fact go to every computer and break there?

Imamichi: The figure in the paper shows that there are optical switches - however the distance involved in the data-way is very long, sometimes greater than 5 km. So a repeater is required to amplify the signal. In practice each computer can have an amplifier, which acts as the repeater - the signal then leaves the computer and goes onto the next station. Optical modems are used to perform this function.

Gellie: You seem to indicate that you load the whole system down from the SCC. How long does it take for the initial loading of the whole system?

Imamichi: In practice we have not actually used down-line loading for the PCs - but it appears in theory that the loading time could take up to 10 minutes. The programs are large - up to 256 kilo-bytes.

Byrne: You say that you have shared-redundancy at both the SSC and DDC level. I would be interested to know how you boot-up the shared devices in the event of a failure of the optical fibre ring? Also, do you have redundancy at the PC level as well?

Imamichi: Yes we have redundancy at the PC level also. Where you have one back-up CPU and 28 PCs and you have a computer failure, the FEPs can detect which CPU has failed. The SSC which has failed is then reloaded under operator control. The FEP has the intelligence to decide which SSC has failed. I would like to emphasize that it is not automatic re-loading - the operator decides to reload by pushing a button on the central control desk.

Gellie: In your paper you showed the curves for 1, 2 and 3 levels of system. It would be intriguing to know where 4 levels would fit on your curves?

Imamichi: When we designed the system we concentrated on the response time and we felt that if we had too many levels the response time of the system would be degraded. Initially we had intended to design a 2 level system but we found that this was not sufficient enough so we went to 3 levels.

Byrne: You mentioned that you have redundancy at the PC level. Could you say if this was in fact shared-redundancy. If so, how did you multiplex all the field I/O?

Imamichi: This is achieved by having a common I/O system, so every PC does not have its own I/O. This implies that the common I/O system is not redundant.

Harrison: Seeing you considered redundancy in so many other places, did you not consider having a redundant I/O system?

Imamichi: We did not intend to have a redundant I/O system because we felt that a fault at that level would be very local. So we considered it unnecessary to provide redundancy here.

Maxwell: I am always fascinated by the speed at which steel mills operate and I wonder whether we can have some idea as to the broad specifications of the plant which he is involved in.

Imamichi: The total length of the plant is about 800m and the sheet being rolled is extracted from the furnace every 60 seconds. At times there are 5 slabs on the line and the minimum gap time is about 6 seconds. This implies that the speed of the slab at the end of the coil could be up to 120 km per hour.

A MESSAGE BASED DCCS

H. Kopetz, F. Lohnert, W. Merker and G. Pauthner

Institut für Technische Informatik, TU Berlin, Federal Republic of Germany

Abstract: MARS (Maintainable Real Time System) is a project on distributed real time computer control which has as its goal the development of such a system from the point of view of maintainability and reliability in hardware and software. This paper presents the architecture of MARS and introduces by a number of examples the programming primitives of MARS. The facilities for reliability and functional enhancement of a MARS system, as well as the possibility for the dynamic reintegration of repaired components are also explained.

Keywords: Maintenance, Reliability, Real Time Systems, Embedded Systems, Distributed Systems, Interprocess Communication, Redundancy

This work has been supported by the German Ministry of Research and Technology (BMFT) under Research Contract IT 1018.

1. INTRODUCTION

The high costs for the maintenance of real time systems is a subject of widespread concern. According to /DeRo 78/ the cost of keeping a successful real time system in a state which is relevant to its users considerably surpasses its initial development cost. These maintenance costs are caused by

- Repair: Repair is necessary because the hardware fails and the design (software) contains design faults which have not been detected during the careful preinstallation tests.
- Functional enhancement: A successful system changes the environment which again changes the requirements for this system. In order to keep a system in a state which is relevant to its users it is necessary to repeatedly modify and enhance the system functions.

We feel that these maintenance activities should not be dealt with only after the system is successfully installed but should be a determining factor during the system design.

The MARS (MAintainable Real Time System) project has set as its objective the design of an architecture and the prototype implementation of a Distributed Real Time System for process control applications from the point of view of maintenance and fault tolerance. In contrast to other projects on fault tolerant architectures /Wens 78,Hopk 78/ MARS is based on the availability of self-checking components.

System Architecture

Any MARS System /Kope 82/ can be decomposed in a cluster and the environment (of the cluster). The environment can consist of the physical equipment which is to be controlled (plant) and/or other clusters. Thus from the point of view of any cluster the rest of the system is considered to form the environment. The interconnection between a cluster and its environment is realized by interface components. The interface components translate the standard information representation in MARS to the form required by the environment.

A cluster consists of a set of components interconnected by an intercluster communication system. A component is a self-contained computer. All components have access to the global physical time. The intercluster communication system can transport messages between the components. A component consists of the MARS machine - that is the computer hardware and executive software -, and the application software, called a MARS module. A module consists of a set of parallel tasks including a priority task which has the authority to reset any of the other tasks.

2. INTERPROCESS COMMUNICATION

2.1 Event and State Messages

In the analysis of real time systems it is helpful to distinguish clearly between event and state information. Event information deals with the occurrence of events (an event is a happening at a point in time). State information deals with attribute values of objects which are valid for a given time interval. State and event information are closely related - the change of an attribute value of an object is an event. In order to facilitate the exchange of event and state information between tasks, the concepts of event and state messages are introduced. Depending on the point of view taken by a receiver, a given message can be handled by one receiver as a state message, but by another receiver as an event message.

The difference between event and state messages concerns the handling of these two message classes at the receiver's end. The handling of event messages conforms with the "classical" message semantics. An event message is queued at the receiver when it arrives and dequeued when it is read.

There are two real time values included in every message, the send time and the validity time.

We assume that there exists a global real time base in MARS /Lamp 78/. In an event message the send time denotes the time of event occurrence while the validity time determines upto which point in time this event message is relevant. After this time the event message is discarded by the communication system.

State messages refer to attribute values of an object which are valid in the interval (send time, validity time). State messages are not queued at the receiver. On reading, state messages are not consumed, i.e. the valid version of a state message can be read several times.

In MARS every state message has to be further classified as either consistent or immediate. This distinction is based on the point in time when a new version of a state message will replace the present (valid) version. If a state message is declared "consistent", a new version will replace a valid version only after this valid version has lost its validity. If a state message is declared "immediate" a new version will immediatly update the present version of the state message.

In a distributed real time system it is sometimes not possible to achieve consistency and speed simultaneously. It is up to the programmer to make a decision in a given application environment.

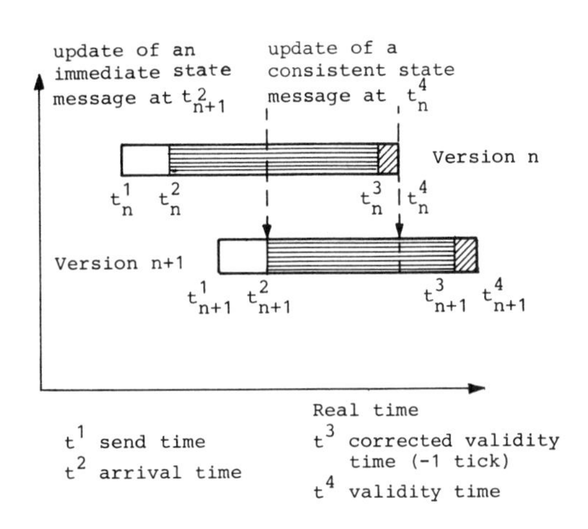

Figure 2.1 Difference between consistent and immediate state messages

The synchronization between sender and receiver is different for event and state message exchanges. Not considering exceptional conditions, the receiver has to read every event message sent by the sender and vice versa, i.e. sender and receiver must operate at the same rate (tight synchronization). In a state message exchange this requirement is relaxed, i.e. sender and receiver can operate at different rates (loose synchronization). It is our opinion that this loose synchronization is sufficient for many information exchanges in process control systems, resulting in simpler interfaces and increasing the autonomy of tasks.

2.2 Data Field

In Real Time Control Systems many tasks are executed cyclically. For example, a control algorithm periodically reads the measured values and outputs the required valve settings or an alarm task periodically scans the measured variables and tests them for alarm conditions. We consider it as an unnecessary restriction to require that the cycles of all these periodic tasks have to be synchronized tightly. This is the reason why we introduced the concept of a cluster "data field", i.e. a set of consistent state messages. The data field is a distributed real time representation of the state of the cluster environment.

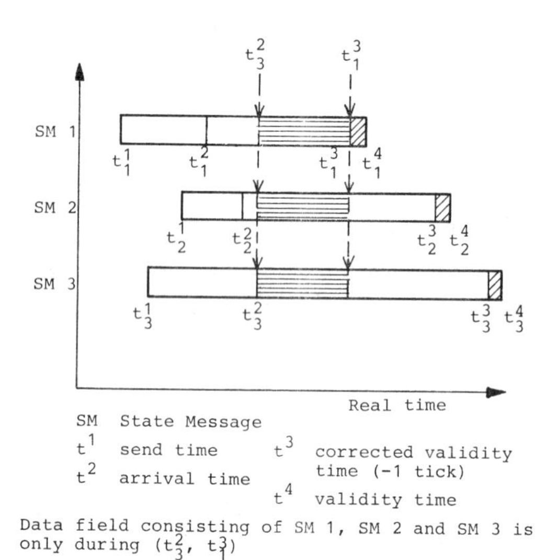

Figure 2.2 Consistency of the Data Field

At any point in time a variable of the data field either contains a value which refers to a valid and consistent (in relation to real time and version) attribute of an entity of the environment or it is undefined (Fig 2.2). The data field is constructed with the aid of periodic state messages. The problem of establishing a valid and consistent data field is non trivial. At present some further investigations are being carried out in order to develop protocols which will establish such a valid and consistent data field. The results of this analysis will be used to determine the time parameters for the state messages, e.g. at what time should a state message become invalid. They will be used in the MARS software development system to support the design of real time programs.

Considering the semantics of the consistent state messages, any component can read the data from the data field with loose synchronization, e.g. if a plant operator wants to look at a certain plant variable he has to access the corresponding element in the data field and can update his display according to his requirements.

The size of the data field is delimited by the channel capacity of the transmission medium and the frequency of the updates of the data field. Fig. 2.3 gives an estimate of the size of the data field.

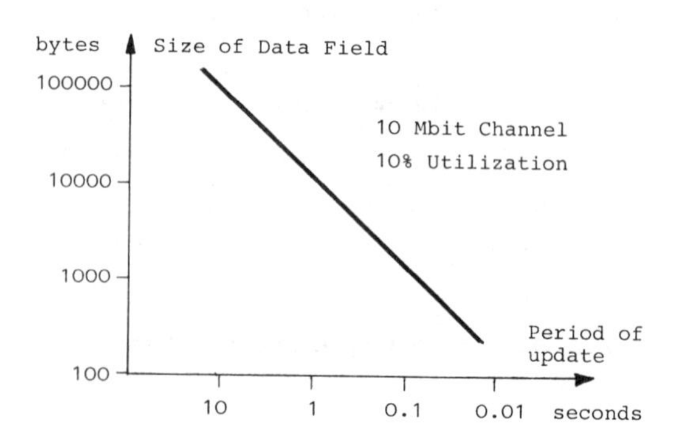

Figure 2.3 Estimation of the size of the Data Field

The detection of errors of the data field (e.g. missing messages) is performed by an autonomous maintenance component.

A component which can read and write from/to the data field is called an active component (in relation to this cluster). A component which is only allowed to read data from the data field but does not have the authority to write into the data field is called a passive component. Because of the naming conventions a passive component can be added to a cluster without any interference with the function of the cluster. This characteristic of the MARS Architecture is very useful for testing purposes. Any operational cluster can serve as a testbed for a new component.

3. PROGRAMMING

The MARS-Project is not only concerned with the development of an architecture for a MAintainable-Real-time -System, but also with the design of appropriate language constructs and a prototype implementation. For this purpose we decided to extend an available programming language. We have chosen PASCAL as a base language for programming the MARS tasks.

3.1 Declarations

Module declarations:

As mentioned above, the application software of a component is called a module. Since the function of a component is determined by the application software, the module identification consists of a function name and a version name. The MARS system supports a one-to-many communication pattern. Modules can be members of one or more groups.

```
MODULE function.version
              MEMBER_OF group;
  .
  .
END function.version.
```

Each message addressed to a particular group will be delivered to all modules in this group. The sender of a message does not know the number of receivers.

External message declaration:

Every MARS message consists of a predefined message header and an optional user defined record. The message header contains the following information:

- name; the message name
- version; the sending module/task version name
- sender; the sending module/task function name
- receiver; the name/groupname of the receiving module/task
- sent; the send time
- valid; the validity time

At the beginning of a module -the module header- all messages which can be received and generated by this module from/to other modules have to be declared. We call the messages, which are declared in the module header, external messages.

```
MODULE function_name;
   IMPORT inmsg  = RECORD ... END;
   EXPORT outmsg;
      .
      .

END function_name.
```

The module 'function_name' receives all messages with the name 'inmsg' with a user defined record and it generates messages named 'outmsg'. This 'outmsg' contains only the message header, i.e. it is a pure signal message.

Task declaration:

As mentioned before, every module consists of a set of parallel tasks. In the header of each task, all messages which can be received and generated by this task have to be declared. If the messages remain within the component, we call them internal messages. Internal messages are only declared in the task header, but not in the module header.

Internal message declaration:

Each internal message declaration declares a task local variable of the same type as the message. In addition to the message attribute EXPORT, input messages have to be classified as either STATE or EVENT messages. Furthermore an attribute INTERVAL must be specified for EVENT messages. This interval declaration is used for the detection of redundant event messages. This interval is opened with the arrival of a message. The interval is closed after the specified duration. All but one of the messages which arrived within this interval are considered redundant and only a single message is delivered to the application software. A message which arrives after the closing of an interval will open a new one. The state and event semantics are explained further in section 2.1.

A successful input operation assigns the actual message, including the header, to a local message variable with the same name as the message.

```
MODULE function_name;
   IMPORT push_button
                    = RECORD ... END;
   IMPORT position
                    = RECORD ... END;
   EXPORT action
                    = RECORD ... END;
      .
   TASK task_name;
      STATE position
       = RECORD ... END IMMEDIATE;
      EVENT push_button
                 = RECORD ... END
                   INTERVAL duration;
      EXPORT action
                    = RECORD ... END;
      .
   END task_name;
      .
END function_name.
```

This module receives one or more

messages 'push_button' and 'position'. The task 'task_name' declares the 'position' as an immediate state message and with an input operation the 'position' is assigned to a local variable 'position'. The message 'push_button' is treated as an event and one or more messages 'push_button' which occur during the interval 'duration' are considered as a single event. In addition the task sends a message 'action' to another task.

4. INPUT and OUTPUT Statement

Because the semantics of the interprocess communication statements in MARS differ from the semantics normally associated with send and receive statements, two new keywords, INPUT and OUTPUT, are used in MARS.

4.1 The OUTPUT Statement

The OUTPUT statement is used to output a message to the communication system. The issuing task then proceeds (i.e. the issuing task performs a rendezvous with the local part of the communication system). The OUTPUT statement requires the specification of a validity time for the message.

```
OUTPUT msg TO receiver VALID time;
```

- 'msg' is a declared message that will be transmitted to the
- 'receiver', that is a component name, a component group name, a task name, or a task group name.
- 'time' is a time expression (relative or absolute). After this time the message will be discarded by the communication system.

At execution of the OUTPUT statement the message header is constructed from this information. The send time and the sender name are inserted into the header by the MARS machine.

4.2 The INPUT-Statement

With the INPUT statement a message can be read from the communication system into a task local message variable. It is a 'selective receive' - from all queued messages the denoted one(s) will be selected.

The general form is:

```
INPUT
      msg11 FILTER f11
      AND
      msg12 FILTER f12
      AND ...
         => stmt1
   OR
      msg21 ...
         => stmt2
     .
     .
     .
   AFTER time => stmta
END
```

- 'msgij' is a declared message that should be received

- 'fij' denotes a predicate on the full contents of 'msgij', (i.e. the message header and the user-defined record), and task local variables. 'msgij FILTER fij' means that only if the evaluation of 'fij' delivers 'true' this message will be read.

- 'msgij ... AND msgik ...' determines that receiving is only possible if all listed messages are available.

- the keyword 'OR' denotes message reception alternatives.

- if no message(s) can be read until a specified point in time, the time out alternative ('AFTER time') is executed.

- if receiving is possible the selected message(s) will be
  - for an event message consumed and assigned
  - for a state message only assigned

  to the local message variable specified in the declaration. The following statement 'stmti' will then be executed.

5. EXAMPLES

Consider a conveyer which is used to transport workpieces. The workpieces have to be counted and sorted according to size (Fig 5.1).

The state of this 'plant' can described by the following information:

- total number of processed workpieces
- number of large workpieces in collection bin
- lever position
- workpiece in process (i.e. a workpiece has interrupted the lightbeam and is not in a collection bin)

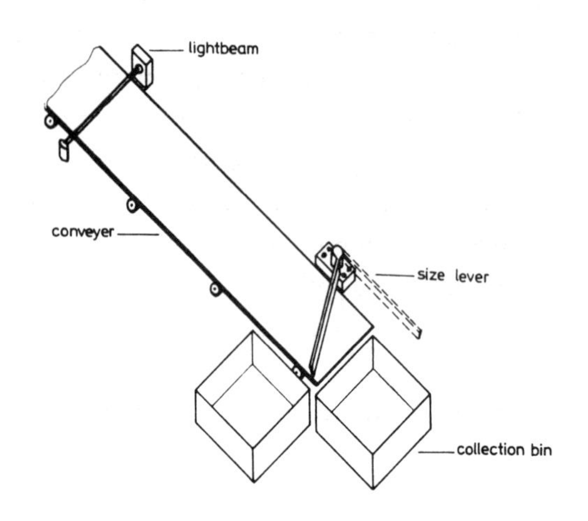

Figure 5.1

Let us assume a constant speed of the conveyer. The following times are relevant.

- succession;	the minimal time between two workpieces
- switch;	the time needed to switch the lever
- transport;	the time needed to transport a workpiece from the lightbeam to the lever

It is required that the conditions

succession > transport
and
switch << transport

are fulfilled, i.e. at most one workpiece may be between the lightbeam and the conveyer end.

The controlling software has the

following general structure (all tasks are implemented as 'loop forever'):

```
TASK lightbeam;
EXPORT workpiece
  = RECORD size : (large,small) END;
.
wait_for_workpiece;
determine_size_of_workpiece;
OUTPUT workpiece TO
    piece_processor VALID transport;
.
END lightbeam;
```

The task 'lightbeam' produces a message 'workpiece' if the lightbeam is interrupted. The contents of this message describe the size of the workpiece.

```
TASK count MEMBER OF piece_processor;
EVENT workpiece INTERVAL jitter;
EXPORT total_number = RECORD ... END;
.
INPUT workpiece => management_info
AFTER succession SEC;
OUTPUT total_number TO  ...
.
END count;
```

The task 'count' processes the information 'workpiece'. Furthermore, it sends at least all successive seconds its internal state, i.e. the content of the message total_number.

The predicate

    jitter << succession

is always true. This must be guaranteed by the plant process.

```
TASK size_lever_position
     MEMBER OF piece_processor;
EVENT workpiece INTERVAL jitter;
.
INPUT workpiece
    => if new_pos then switch_lever
AFTER succession SEC;
OUTPUT size_lever TO ...
.
END size_lever_position;
```

The task 'size_lever_position' changes the lever position if necessary.

Functional enhancement:

Very often after the installation of a system new requirements have to be implemented. As a third selection criteria the weight of the workpieces has to be considered. A weighbridge and a second sorting lever must be installed.

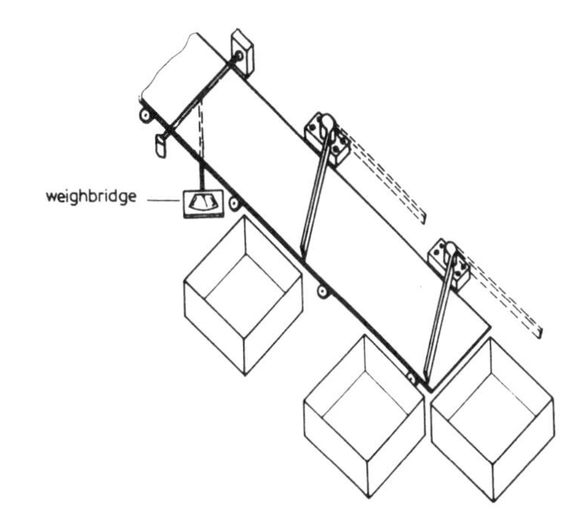

Figure 5.2

This functional enhancement requires the following software additions:

```
TASK weighbridge;
.
INPUT workpiece => ...
AFTER succession SEC;
OUTPUT weight TO ...
.
END weighbridge;
```

```
TASK weight_processing;
.
INPUT weight => ...
AFTER succession SEC;
OUTPUT weight_lever TO ...
.
END weight_processing;
```

Both tasks can be added to the system without interfering with the existing system.

Reliability improvement:

Let us assume, that it has been shown in practical operations that the lightbeam is a reliability bottleneck. A second, i.e. a redundant lightbeam has to be installed. An additional task 'lightbeam redundant' (only a copy of 'lightbeam') has to be included.

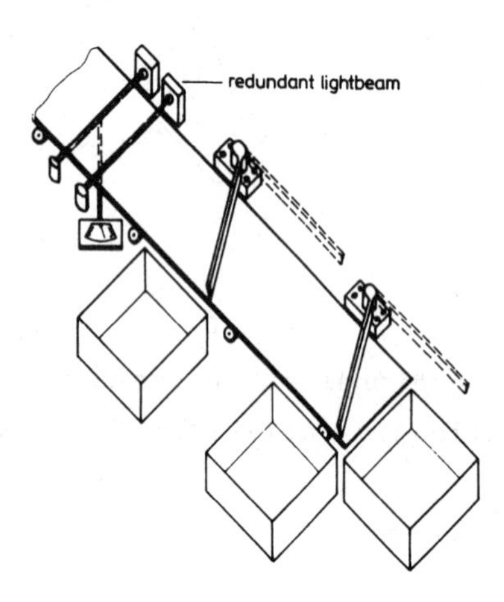

Figure 5.3

The addition of this redundant task is simply a 'plug in'. The interval semantics on the receivers side of the message(s) 'workpiece' supports the insertion of multiple 'workpiece' senders.

Under the assumption that the lightbeam is self-checking a failing lightbeam can be detected by a maintenance task which accepts all 'workpiece' messages by a simple event declaration and checks for completeness of the messages 'workpiece'. This support of multiple viewpoints (function, error detection) by the MARS architecture reduces the complexity of the solution.

The addition of a task with only functional behavior (i.e. the produced output at time t is independent from the work done before) is easily achieved. Tasks with an internal state can also be added with the following scheme:

```
TASK one_of_multiple_copies
          MEMBER OF copies;
STATE internal
  = RECORD my_state : structure END;
.
.
{ initialisation }
INPUT internal => { recovery start }
  AFTER (cycle + trans) SEC
        => { initialisation start };
.
.
{ loop forever }
INPUT request => perform_service
AFTER cycle SEC;
OUTPUT internal TO
                  copies VALID cycle;
.
END one_of_multiple_copies;
```

In this scheme the task 'one_of_multiple_copies' produces after each request a message 'internal'. This message holds the internal state information of this task. Furthermore, the task outputs its internal state each 'cycle' seconds. The restart of redundant tasks is therefore independent of the rate of requests. In the initialisation phase a new task waits on a message 'internal'. After receiving such a message the internal state of the other (active) task is known. If no message 'internal' can be received within the duration cycle + trans then it must be assumed that this task is the only one in the cluster producing a message 'internal'.

However, this assumption is valid only if the following condition holds

restart > 2 (cycle + trans)

- restart; the minimum duration between the start of two redundant tasks.
- cycle; the maximum duration between the sending of two messages 'internal'.
- trans; the maximum time to transport a message between the redundant tasks.

Until repair of a failed component of the system a redundant component is still available .

Mapping the task to computers:

Because in MARS there is no distinction between external (module-module) and internal (task-task) communication all tasks can be freely mapped to MARS machines (i.e. hardware). This mapping has to be done in such a way that the overall reliability becomes a maximum.

6. LITERATURE

/DeRo 78/ De Roze, B.C., Nyman, T.H.: The software life cycle - a management and technological challenge within the department of defense, IEEE Trans. SE-4, July 1978, p. 309-318

/Hopk 78/ Hopkins, A.L., Smith, T.B., Lala, J.H., FTMP - A highly reliable Fault Tolerant Multiprocessor For Aircraft, Proc. IEEE, Vol. 66, No. 10, Oct. 78, p. 1146-1154

/Kope 82/ Kopetz, H., Lohnert, F., Merker, W., Pauthner, G., The Architecture of MARS, Report MA 82/2, TU Berlin, April 1982

/Kram 83/ Kramer, J., Magee, J., Slowman, M., Lister, A., CONIC: an integrated approach to distributed computer control systems, IEE PROC., Vol. 130, Pt.E, No. 1, January 83, p. 1-10

/Lamp 78/ Lamport, L., Time, clocks and the ordering of events in a distributed system, CACM, Vol. 21, No. 7, July 1978, p. 558-565

/LeLa 79/ Le Lann, G., An Analysis of Different Approaches to Distributed Computing, Proc. 1st International Conference on Distributed Computing, Huntsville, October 1979, p. 222-232

/Svob 79/ Svoboda, L., Reliability issues in distributed information processing systems, Proc. of the 9th International Symposium on Fault Tolerant Computing, June 1979, p. 9-16

/Wens 78/ Wensley, J.H., Lamport, L., Goldberg, J., Green, M.W., Levitt, K.N., Melliar-Smith, P.M., Shostak, R.E., Weinstock, C.B., SIFT: The Design and Analysis of a Fault-Tolerant Computer for Aircraft Control, Proceedings of the IEEE, Vol. 66, No. 10, October 1978, p. 1240-1254

DISCUSSION

LaLive d'Epinay: I would like to know how Professor Kopetz copes with the problem of the equivalence interval being larger than the repetition interval of an event?

Kopetz: If this is the case the system will not operate! The system, during design, should show that you have a problem which you have to solve in some other way. We found that when we looked at a variety of applications we never ran into this situation, but if it had arisen our approach would not work.

Plessmann: It would be very useful and interesting to know how the equivalence interval should be specified.

Kopetz: It is in effect an application dependent time interval which must be extracted from the application. If you refer to the paper you will see that in the example of the conveyor belt which is counting packages travelling on the belt, you might not be happy with the reliability of the photo cell, so you could simply put in a second one. The equivalence interval is now determined by the minimum distance which can exist between any two packages, and the speed of the conveyor belt. The time that determines the distance between two packages will be the parameter used for the equivalence interval and any event message which arises within that equivalence interval is considered to be a redundant copy of the previous one.

Rubin: I would like to hear more information from Professor Kopetz as to the exact redundancy scheme which his architecture employs.

Kopetz: The redundancy which can be implemented in the MARS architecture may be active redundancy or stand-by redundancy. The architecture will support, without any change in the application software, active redundancy in the form of self-checking components. By active redundancy we mean that both components are operating actively in parallel. If we are not happy about the reliability of a component and wish to include redundancy, we just plug in a second component with the same software. This means that we are only catering for degradation-type faults in the hardware. The addressing scheme used will give the inputs to the other components and the state message semantics will remove the duplicate messages at the output side. If one of the components fails the other one can take over. The second part of the problem revolves around the re-integration of the repaired component. Each component has the same self-checking property which means that, if the component does not do what it is supposed to do, it will shut itself off - that is the property which is involved in the checking. This implies that part of the duty cycle of the component is self-testing.

Rubin: How do we then cope with the problem of sensors and actuators which are not inherently self-checking?

Kopetz: I am talking about computing functions in a computer system and the question is how to get an image from the plant which is consistent. Say you had a process in which you had three temperature sensors and each one measures the same temperature, but with a slight variation. In order to resolve this, I would consider that you need an application dependent module which would read all three temperatures and do statistical analysis of them, and then output the best estimate of the temperature into the system. If you want to make this system redundant you can do it by having two components performing this analysis. Clearly the self-checking components would each produce a message indicating the temperature. These messages would be defined as consistent state-messages which would not lose their validity until the expiry of the validity time.

EXPERIENCE WITH A HIGH ORDER PROGRAMMING LANGUAGE ON THE DEVELOPMENT OF THE NOVA DISTRIBUTED CONTROL SYSTEM*

F. W. Holloway, G. J. Suski and J. M. Duffy

Lawrence Livermore National Laboratory, University of California, P.O. Box 5508, Livermore, California 94550, USA

Abstract. Interest in the impact of high order languages (HOL) on real time and process control applications development has intensified recently, due in part to the Ada language development effort sponsored by the U.S. Department of Defense. High order languages, such as Ada, incorporate state-of-the-art features in the areas of modularity, data abstraction, separate module compilation, strong type checking of data, multi-tasking, distributed processing and exception handling. The intent of such languages is to improve programmer productivity, software maintainability, and the effective management of software development in large real time systems.

This paper explores the impact of an HOL on the development of the distributed computer control system for a large laser fusion facility called Nova. Nova utilizes fifty microcomputers and four VAX-11/780's in its distributed process control computer system.

Praxis, a high level real time computer language, was designed and implemented for Nova to support the control system development activity. Praxis is similar to Ada in its design goals, has many features comparable to those found in Ada, and has had operational compilers for over two years.

The purpose of this paper is to assist readers who are beginning to evaluate the importance of HOL's in small to medium scale process control applications. It includes a summary of the major design objectives and features of these HOL's which is followed by a description of our experiences with Praxis. The article concludes with our preliminary expectations and concerns for the use of Ada within process control applications, based upon Praxis experience.

Keywords. Ada, Distributed Control Systems, High Order Language, Software Management.

*Work performed under the auspices of the U.S. Department of Energy by the Lawrence Livermore National Laboratory under Contract No. W-7405-ENG-48.

INTRODUCTION

Many assert that the dominant programming language of the next decade is destined to be Ada (Carlson, 1981). However, the compilers for this language are just now becoming available. The culmination of an eight-year effort led by the U.S. Department of Defense, Ada is specifically targeted as a High Order Language (HOL) to meet the requirements of program development and maintenance for embedded real-time control systems. The language offers new and often complex features which are unfamiliar and difficult to evaluate without actual use.

At Lawrence Livermore National Laboratory, a language named Praxis was developed to support the development of software for a fifty computer process control network. This language is similar to ADA in many of its design goals and features. The purpose of this article is to present our experience with Praxis, as a HOL, and our resulting expectations and concerns for Ada in process control applications.

DESCRIPTION OF APPLICATION

Nova's Distributed Computer Based Control System

The application for which our Praxis-based distributed computer control system was developed is named Nova. It is a 150 terawatt (TW), ten arm laser fusion research facility currently under construction at the Lawrence Livermore National Laboratory (LLNL) (Simmons, 1982). As the world's most powerful glass laser system, Nova will provide researchers with an important new tool in the study of inertial confinement fusion (ICF). A principal objective of Nova is to demonstrate the feasibility of generating power from controlled thermonuclear reactions.

An intermediate laser system called Novette (Manes, 1983) utilizing Nova's computer based control system technology, was recently completed at LLNL. Novette utilizes two beams to provide 30 TW of light onto fusion targets. It provides important data on laser operation and target performance in preparation for the Nova experiments, and has proven the Nova control system design.

The Nova system's ten laser beams will be capable of concentrating 100 to 150 kilojoules (kJ) in 3 nanosecond pulses of infrared light onto a fusion target a few millimeters (mm) wide. The system will generate light at shorter pulse lengths in power bursts up to 150 TW. Nova will also generate green and near ultra-violet light by doubling and tripling the fundamental output wavelength of its amplifiers using passive crystal technology.

All ten arms of Nova must deliver their individual pulses to the target simultaneously (within $\pm$ 5 picoseconds). To achieve this objective, a single 100 microjoule pulse is selected and amplified to approximately 50 joules in a single pass through a nine stage preamplifier. It is then split by partially reflecting mirrors into ten parallel chains of power amplifiers, each consisting of fifteen cascaded laser amplifiers. Each pulse emerges from the output of its 180 meter long chain with a beam diameter of 74 centimeters. The pulses are subsequently reflected by large alignment mirrors, converted to shorter wavelengths, and finally focused onto a fusion target inside a 5 meter diameter aluminum vacuum vessel.

Control System Organization

Nova's control system employs a distributed, computer based architecture which evolved from the successful Shiva laser control system (Suski, 1979). It is organized functionally according to four fundamental subsystems; Power Conditioning, Laser Alignment, Laser Diagnostics, and Target Diagnostics. A fifth, unifying subsystem called Central Controls, centralizes, augments, and coordinates the other subsystems' functions. Criteria of reliability and adaptability are met by the computer based, extendable nature of the system. Flexibility required to optimize individual subsystem architectures is provided by the inclusion of the Central Control subsystem. This subsystem establishes a single point at which compatible interfaces for command, control, and data interactions are established. This architecture supports parallel software development in the five distinct areas, with minimal interaction required between groups.

In this hierarchically structured system, approximately fifty Digital Equipment Corporation (DEC) LSI-11/23 microcomputers provide localized control and data acquisition capabilities in geographically distributed locations throughout the laser fusion facility. Data from these front end processors (FEP's) is collected, analyzed and integrated at the Central Control level with three redundant Digital Equipment Corporation VAX-11/780 minicomputers. Remote command and control capabilities, higher level control functions (e.g., automatic laser alignment), and high volume data storage and manipulation, are implemented at this level. All computers are interconnected using either multi-port memories or a specially developed, high speed (10 Mbits/second) fiber optic network.

Seventy-five man years of effort have been expended in developing the Nova control system including high speed network communications, data base management techniques, real time handlers, operator interfaces, and systems and applications level software.

THE PROGRAMMING LANGUAGE ADA

Why Ada

In the mid-1970's, analyses of computer efforts in the United States Department of Defense (DOD) showed that over three billion dollars per year were being expended on software (Fisher, 1978). The majority of this software was used in embedded real time control systems (e.g., aircraft controls). Issues of maintainability and support costs were steering the DOD towards requiring high order languages in all systems. However, the lack of features and efficiency in these languages led to many exceptions to the use of HOL's. Over fifty percent of all real time embedded software in DOD systems was being written in assembly language. This resulted in continuous training problems, logistics problems, and upgrade difficulties.

Consequently, the purpose of the DOD in initiating the Ada effort was to develop a language for real time embedded systems which would be used exclusively on all DOD related projects. The fundamental objectives were:

- o Reduce the cost of software throughout its life cycle
- o Allow development of truly transportable software - both across applications and hardware
- o Allow responsive, timely maintenance of long lived software
- o Support very high reliability needs
- o Increase readability of all software at the (possible) expense of writability
- o Produce high efficiency code, comparable to well coded assembly language.

The History and Current Status of Ada

The Ada language has been under design and development for over eight years. The initial effort was to determine the requirements of a HOL for embedded real-time control systems, and whether existing languages or combinations of these languages could meet these requirements. A series of documents with the code names STRAWMAN, WOODENMAN, and TINMAN were subsequently issued and reviewed by representatives of the academic community, industry, and government. Beginning with STRAWMAN, each successive document became more specific, precise, and complete in stating DOD's HOL design criteria. In 1976, a group of reviewers evaluated existing languages against TINMAN and determined that no available languages would meet the stated criteria. This committee recommended that a new language be designed, and that it be based on either Algol-68, Pascal, or PL/I.

In 1977, four companies were given contracts to design languages to the IRONMAN specification. Their submittals were given code names and distributed widely for review:

Company	Code
CII-Honeywell	Green
Intermetrics	Red
SRI	Yellow
Softech	Blue

All four companies had based their languages on Pascal. Red and Green were selected as the final contenders. A new and final language requirement document, STEELMAN, clarified and corrected inconsistencies in IRONMAN. Red and Green designs were revised and completed against STEELMAN. After extensive public review, Honeywell's Green was chosen. Contracts for language implementation on Digital Equipment VAX and IBM 370 architectures were awarded. These compilers are expected to be completed in the next twelve months.

Additional Ada development efforts around the world have been initiated. These include a compiler by Intel for the Intel 432, the recently announced ROLM compiler, an interpreter at New York University, and several European, Canadian and Asian efforts.

The Ada language is subject to a tightly enforced standard. Softech, Inc. is completing the Ada validation system which will be used to qualify all compilers using the DOD registered trademark, "Ada". This eliminates the tempting possibility of subsetting the language (to ease implementation) while still referring to such implementations as "Ada".

In addition to the Ada language effort it was recognized that, to achieve Ada's objective, the total programming environment must be considered. It was important to provide the programmer with a standard and sufficient set of tools, independent of the host support system specifics. Therefore, a series of consecutive "environment specifications" including PEBBLEMAN (1978) and subsequently STONEMAN (1980) were developed. At least two major efforts are now underway to implement tools such as editors, configuration and database managers, and commonly used run-time support libraries. This work includes the Kernel Ada Programming Support Environment (KAPSE) which presents a virtual host operating system interface to the compiler and other tools. Changes in the actual hardware and operating system can therefore be accommodated within the KAPSE, reducing the need for recoding the support tools for new hosts.

Progress on the environments is not as far advanced as the Ada language itself, yet well-designed environmental tools will be required if Ada is to meet its fundamental design objectives.

The current schedule for Ada indicates that DOD compilers and environments should be available in late 1983 and 1984. Mandating the use of Ada in all new DOD embedded real-time applications is planned by 1990.

Ada Concerns

Despite the exhaustive design efforts and considerable commitment to Ada's success, there are still concerns, and opposition to Ada does exist (Ledgard, 1982, Hoare, 1981). The majority of the concerns treat the two categories listed below:

- LANGUAGE COMPLEXITY. The language is too rich and complex, detracting from its usefulness, reliability, and maintainability. Subsets are highly desirable (though specifically prohibited at this time). A result of this complexity is that non-professional programmers may find Ada too intimidating.

- REAL-TIME RESPONSIVENESS. The lack of known storage requirements at compile time and the possible influencing of its high level task synchronization features may impact the safe, reliable use of Ada. This applies specifically to real-time situations such as flight control. Overhead due to run time checking of ranges or other constraints is also a concern.

Major Features of Ada

Having summarized the motivation and status of the Ada project, we now briefly review selected important features of the language. Also present in the Praxis language, these features are presented in order according to their impact on the Nova control system development effort:

- Extensive Compile and Run Time Checking - All manipulations involving data, types, and other entities are checked for correctness and legality. This includes the parameters used in procedure calls. Run time checks include range checking of all data values.

- Declarations and User-Defined Data Types - The structured type declarations familiar to Pascal users have been extended to improve capability. Ada (and Praxis) require that the types of all data be declared. Types are strictly enforced and only limited coercion of mixed data types is permitted.

- Separate Compilation - Ada allows procedures, variables, types, and collections thereof to be compiled separately as small units, while maintaining strict compile-time type checking across these units.

- Self-Documenting - The length of identifier names allowed, combined with good data structures and excellent control statements, supports writing of self-documenting software. (It is, of course, still possible to write inscrutable code if such capabilities are not well utilized.)

- Extended Control Structures - Extensive control structures are provided to clearly indicate operations being performed. Some are redundant. Praxis does not have the GOTO statement.

- Packages - Ada provides comprehensive methods of grouping procedures, declarations, and collections of data into separately compilable modules. Ada includes library support for packages and generally improves upon Praxis' capabilities.

- Exception Handling - Exception conditions (I/O interrupts, range errors, underflow) can and must be handled in programmer defined high level code.

- Data Abstraction - Type declarations for data and procedures can be compiled separately from the code which uses or comprises them. This unclutters the user interface and prevents unauthorized changes to data or code.

- Enumerated Data Types - Data types may be defined in which values are limited to a list of programmer-defined alphanumeric "names". For example, type color is limited to "values" RED and BLUE.

- Interprocess Communication Constructs - The language provides substantial support for synchronizing processes and interlocking access to shared data.

- Structured Access to Machine Features - Access to machine features is defined and controlled within the language. Frequent use of supporting assembly language programs is unnecessary and discouraged.

Ada includes significant features which were not incorporated into Praxis due to difficulty of implementation or their limited usefulness. These include:

- Generic Procedures - Procedures with identical names but different parameter specifications can be utilized and distinguished by the compiler. This increases the value and flexibility of libraries of preprogrammed procedures.

- I/O - File, text and device oriented I/O is defined as part of the language and implemented in packages.

- Multi-Tasking - This difficult to

implement feature is a major contribution of Ada. Specific constructs for the initiation and control of parallel tasks within the language itself are defined.

- o Standardized Programmer Support Environment - Ada will include standardized tools such as editors, configuration manages, data bases, and user libraries. This feature will also serve to insulate Ada software and tools from changes in operating system software and hardware.

- o Overloading of Operators - Ada allows operators to have different, programmer defined effects based upon the types of the parameters to which they apply.

Figure 1 shows an example of a procedure written in both Ada and Praxis.

THE PROGRAMMING LANGUAGE: PRAXIS

Praxis - Motivation, History and Current Status

We stated at the onset of the Nova Project that substantial savings in time and effort would result if a powerful controls-oriented programming language were available. We had endured several years of dealing with older languages, their restrictions, awkwardness, and inexactness. Extensive debugging sessions often led to discovering a misspelled variable name or a misuse of a variable type. Several different languages and operating system environments had been used since no single product covered the breadth of features needed in this large distributed system. This created typical problems in software maintenance, making support and extension difficult.

Therefore, in January 1979 LLNL issued a contract to Bolt Beranek and Newman (BBN), Inc. to design and to implement a Praxis language compiler for Digital Equipment Corporation PDP-11 computers. The development of Praxis originated from an initial study by BBN, funded by the U.S. Defense Communications Agency (DCA), to determine the requirements of a language for communications programming. With the clarification of the Nova controls hardware architecture and schedule, BBN's work was expanded to include the development of a VAX/VMS native-mode compiler, documentation, additional language design, and a high-level input/output package.

In March 1980 the preliminary PDP-11 compiler successfully passed two critical milestones. The first milestone was that the compiler, which is written in Praxis, compiled itself successfully on the PDP-11 systems, proving that the bulk of the compiler was correctly implemented.

The second milestone was the implementation of a Nova controls application of the language, a ROM-based LSI-11 processor. A 2000-line assembly-language, stepping motor control program was recoded in Praxis, compiled, and burned into read-only memory (ROM). This demonstrated that the language was indeed powerful enough to replace detailed assembly language sequences and that the compiler correctly implemented the controls-oriented features.

In December 1980 we took final delivery from BBN of the completed Praxis compilers, support software, and documentation. The products were:

- o VAX/VMS compiler generating VAX code
- o VAX/VMS compiler generating PDP-11 code
- o PDP-11/RSX-11M compiler generating PDP-11 code and support software and documentation
- o RMS Input/Output package
- o Language Reference Manual (Evans, 1981) (300 pages)
- o Input/Output Manual
- o Compiler Internals document

In addition we completed two in-house documents:

- o An Introduction to Praxis
- o Programming in Praxis

We have been using the Praxis language for control systems programming since the Summer of 1980 with remarkable success and acceptance. More than 300,000 lines of operational Praxis code have been written.

The Praxis language is specifically within the state of the art of language design. It was particularly designed for control and system implementation needs. It is a comprehensive, strongly typed, block-structured language in the tradition of Pascal, with much of the power of the forthcoming Ada language. It supports the development of systems composed of separately compiled modules, user-defined data types, exception handling, detailed control mechanisms, and encapsulated data and routines. Direct access to machine facilities, bit manipulation, and interlocked critical regions are provided within the language.

Since the control system environment differs in important ways from application to application and machine to machine, Praxis has features to handle these differences. High-level facilities that mask machine dependencies and foster portability often restrict real-time, control applications programming. However, Praxis is a high-level language that has controlled access to machine dependencies.

Complex language features, such as Ada's generic procedures, overloading of operators, and parallel processes, have been intentionally left out. We felt that these concepts were either not understood well-enough to be incorporated at this time, or that they need not be part of the language.

Summarizing, Praxis is a powerful, modern programming language that goes beyond Pascal and has been used for over two years. At this time it would be difficult to prove or disprove any cost savings due to the use of Praxis, but a preliminary version of the Nova control system is now operational in the smaller scale version of Nova, Novette, and the system operators are satisfied. The great majority of all of the software for Novette was written in Praxis and the writers are satisfied. Furthermore, we have not found any application where Praxis was found to be insufficient.

HOL USE IN A DCCS

Experiences with the Praxis HOL

The following information on the actual use of Praxis is based upon personal experience, formal interviews with Nova project programmers, statistical analysis of the 300,000 lines of Praxis source code, and informal communication between all of the Praxis users at LLNL. In presenting the use of these features, we contrast the frequently used and popular ones with those that are impractical, difficult to understand or seldom used. Areas where additional needs remain are also discussed.

Frequently Used Operations in Nova

Predictably, the integer and bit data declarations, and the more traditional flow control operations, were the most frequently utilized statements. Note, however, that while neither Praxis nor Ada contain true string operations, the number of ASCII strings encountered within the Nova control system software was surprisingly large. Approximately 19% of all source code lines contained ASCII strings. A significant portion of these arise from debugging messages in the code. However, a significant portion is also found in software supporting man-machine interactions through operator consoles and other peripheral devices. There should not be any more doubts concerning the importance of string handling operations within control systems. Ada's generic capabilities support the implementation of string operations, and we urge that a standard emerge from the work on environments.

Features and Characteristics Most Liked in Praxis

The availability of Praxis' extensive data structuring facilities was welcomed enthusiastically by our control system implementers. In retrospect, we over utilized this feature and now are retreating. The occurrence of complex uses of large central data structures in software products actually detracts from maintainability. They constrain advances and are difficult to design for long term benefit.

The single most valuable characteristic of the Praxis language has been the relative completeness with which the compiler can check that the code represents the intent of the programmer. Our experience indicates that once a program has successfully compiled, it will run as expected with little or no machine interactive debugging.

The key features of the language that make this possible are compulsory declarations, enumerations, and tightly enforced type checking. Separate compilation of modules is also a practical requirement.

Large identifier names that are meaningful to the application were judged to be very valuable. It is interesting that this single feature, which is simply and easily implemented in any language, ranks near the top of the list in value to the users. However, for those about to buy programmer workstations, be aware that typical statements in these languages are long. Identifiers in Praxis have a 32 character limit, often used in interests of readability. With block indenting and multi-level structure references, even 132 character wide terminals can be limiting. As a further (minor) comment, the standard eight character wide tab spacing, used by several manufacturers, is too wide and awkward for this type of language.

Features Found Impractical for Nova

Impractical features are those which cost more to design, implement or utilize than will be returned in benefit. For example, the designers of Ada and Praxis have attempted to create languages that totally replace the need for native assembly language. At least with Praxis, it has been proven to our satisfaction that the use of pseudo in-line assembly code is impractical. The effort to implement this feature and educate users on methods and restrictions was, in retrospect, of little net benefit. With the exception of one individual, in-line code was seldomly used. It was never been found to be essential.

Another impractical effort was developing a machine code generator in lieu of assembly code. The Praxis compiler was originally written to produce assembly language statements which were assembled by a standard assembler. This provided an extremely valuable debugging tool - the intermediate assembly language code - which was appreciated by the users. The mysteries of just what the compilers did with exotic source statements, and bugs related to the interrelationship of Praxis code with its environment, were often resolved through close scrutiny of the assembly language code produced by the compiler. However, in the interest of improving overall compilation speed, this feature was removed from the compiler that generated VAX code. Fortunately, it remained in the LSI-11 code generating version, where it was most useful.

Yet another possibly impractical, but enjoyed, feature of Praxis allowed a carriage return to be a statement terminator instead of a semicolon. This contributed to productivity and readability. However, it used some rather complex rules to determine when end-of-line was actually end-of-statement. As a result, it consumed a considerable amount of time and effort in Praxis compiler design and implementation that may have been better spent testing other features.

Aspects of Praxis and Ada Which Are Difficult to Understand

Ada and Praxis are strongly typed languages. Every use of every identifier is checked for consistency with the original specification of its type. This specification is often in another module (package). This introduces a major complexity in organizing the location of type specifications and in specifying the order in which modules may be compiled.

Frequently, adding a reference to an object used by other modules to a module being modified causes the established order of compiling modules to fail. Often, for seemingly simple functional changes, large structural modifications must be made to the location of type definitions, leading to massive editing efforts.

It is frequently difficult to structure software to meet the requirements introduced by separate compilation and still maintain the flexibility required to easily adapt to changing application requirements. One user has suggested automatic generation of an ordered list of all modules required to be compiled prior to compilation of a given module. One aim of the Ada environment efforts is to treat this specific problem in an automatic or semi-automatic manner.

Seldom Used Features of Praxis

The ability to combine source statements on a single line separated by semicolons was seldom used. Since Praxis source lines tend to be lengthy, due to long identifier names and indentation conventions, there are few cases where multiple statements on a single line are desirable. In fact, process control programmers often prefer writing a series of vertically arranged "steps". With the exception of one individual, only 0.6 percent of all source lines contained multiple statements in Nova.

Features providing assembly level capabilities were often found to be less useful in the HOL. For example, the exclusive OR operation, included in many languages, was used exactly three times in the entire system. Another case is the clever SWAP operation where the contents of two variables are exchanged. Often used in assembly level programs in prior systems, SWAP has never been used in Nova.

More importantly, we note that redundant, complicated methods of advancing indices on loops are often not used. A single, simple technique is often selected and employed exclusively.

Features Desired but Not Implemented

The Ada language was designed with three overriding concerns; program reliability and maintenance; concern for programming as a human activity; and efficiency. The need for languages that promote reliability and simplify maintenance is well established. Hence, emphasis was placed on program readability over writeability (Morgan, 1983).

Concern for the human programmer was stressed during the design. But what about the software manager? Praxis did little to assist us with what is the remaining most difficult part of software development, its management.

In a loosely organized process control system development effort, a set of compiler controls is required to restrict the use of features which conflict with long-term objectives or are otherwise determined to be undesirable in a particular application. Also of use would be enforced entry of certain information, such as author, project name, date and revision history, organization title and copyright notice. Better debugging capabilities are needed for Praxis and planned for Ada. In Nova, debug messages are commonly inserted at procedure entry and exit points to show the state of parameters. A compiler should generate these and other messages at designated points according to directives. Higher speed compilation and linking with more aids to regenerating large software products are also desirable.

Out of our interview process came a request for a simple initialization operation which could be applied to any data structure. Also requested was that all declared variables default to some known condition. The ability to express the type of storage allocation allowed on parameters to a procedure (for example, that a parameter must be in I/O space of memory) is desirable. Users would like to see FORTRAN formatted I/O incorporated. This is not surprising given that I/O in Praxis was not part of the language definition.

We note that many of these comments are applicable to Ada. However, the Ada language definition is fixed and not likely to change soon. Some of these concerns can and are being addressed in the work on Ada environments.

FEATURE USE WITHIN PRAXIS

Statistical Survey

Our Nova control system effort provides a reasonable size sample set for examining the actual use of features within a high order control system language. Several hundred thousand lines of code were written by approximately twenty-five people over a three year time period.

A statistical survey of the use of Praxis features within the Nova control system software, including the Praxis compiler, was performed. The incidences of every reserved word and symbol within the language, user defined symbols and types, sizes of modules and lines, and compiler directives were measured. Also recorded was each use of general system software packages. This data was correlated with several categories of users.

Programming Cultures

From correlations, we determined that user groups often did not use the features which they had originally requested. More significantly, clear dialects of usage have developed within small teams. This is taken as evidence that even Praxis, which is sparser than Ada, has redundant features. Following the lead of Barnes (Barnes, 1980) we correlated the use of features with several "cultures" within the group of developers:

Compiler writers - This culture consists of the professionals who specifically design and write language compilers.

Professionals - The professional culture is concerned with writing programs of a permanent nature. These programs are usually large. They are written by teams of individuals whose profession is primarily the design and writing of programs. Their programs should be adequately documented to accommodate maintenance over a several year lifetime. Within this group, it is essential that the language used be standard, stable and reasonably well-known. Important characteristics are the need for separate compilation, readability and compile time error detection. Interactive use is not required. Examples of existing languages used by this culture are FORTRAN, Pascal, COBOL, PL/I, CORAL 66, RTL/2 and CHILL.

Amateurs - The amateur culture is concerned with writing programs of a less permanent nature. Individuals in this category have primary professions in different fields such as accountancy, medicine, or chemical engineering. They use the computer as a tool and their programs are often used only by themselves. There is consequently little need for maintenance and documentation. Important characteristics of languages for this group are ease of writing and general "user friendliness"; interactive use is desirable. Examples of such languages are BASIC, FORTH, APL, and the command language of many operating systems.

Novices - People in this category are just beginning to learn about languages and the application of computers.

Table 2 lists overall statistics according to these cultures. Within this summary data some interesting trends are evident. On a percentage basis, the amateurs documented their code with comments substantially more than the compiler writers and professional programmers (35% versus 25%). The professionals wrote substantially fewer assignment statements and had more module interface declarations, procedures, and type statements. They made powerful use of aggregate assignments and procedure calls. The professionals tended, surprisingly, to write larger modules with more complex interfaces. Amateurs more frequently ventured into unusual features. Experience showed that this caused them some difficulties.

Of note is the total of over 1/2 million references to identifiers (variables, constants, procedures, functions) required to control the Nova system. This emphasizes the advantage to be gained by having strong compile time checking of each such reference. Programmers employed user-defined data structures freely. Over 4,000 user-defined types were employed (see "IS" in Table 3). Other trends can be noted from the data. For example, there is a clear difference in the use of OPTIONAL (parameters to procedures and functions), INITIALLY (initial conditions), and EXCEPTIONS by the various cultures.

PROJECTIONS AND CONCLUSION

Projections and Cautions

Our experience with Praxis on the Nova control system indicates that the content and features of the Ada language are outstanding. We perceive that the present implementation efforts may be getting ahead of overall environmental considerations. Such environmental features as improved compiler and linker speed, interactive capabilities, ease of writing, and general "user friendliness" should be emphasized. In Nova, the single greatest contribution to programming style and productivity was Praxis in conjunction with a user-oriented screen text editor.

Despite our predictable startup problems with the brand new Praxis language, all of the users reported satisfaction with the overall impact of the language on their projects. Moreover, they intend to use the language, or Ada if it is available, on their next project if possible.

Ada and to a lesser extent Praxis, are languages rich in syntax and features. There are a sufficient number of redundant features in Ada that dialects will develop. The result in large systems can be that portions of the system written by one team or individual will not be easily understood by other teams or individuals. While sharing of such dialects can be a stimulant to learn, we urge the use of standards of programming style. Such standards should limit excessive proliferation of dialects in applications which will require long term maintenance.

Conclusion

We feel that Ada, with good environmental standards, will provide an intellectual stimulant to advancing professionalism in software development for distributed computer control systems. However, in planning to use Ada one should note that the language by itself will not ensure better, more maintainable control systems. Nor will first time use of Ada on small to medium size systems aid in achieving their timely implementation. The impact of factors such as compiler speed, a rich syntax, strong typing, and separate compilation on initial programmer productivity should be anticipated and factored into project plans.

Software written in such a language, which compiles successfully, often executes successfully the first time. This single characteristic may have a significant impact, in the long term, over the way programs are developed and future languages are designed. Programmers may no longer need to spend most of their development time on the actual control systems. The availability of good, centralized host development facilities can be used to bring developers into closer proximity. The resulting improvement in natural communication can produce higher quality computer control systems in distributed applications such as Nova.

Acknowledgments. We wish to acknowledge the assertiveness and assistance of James Greenwood (former LLNL employee) without whose drive Praxis would not have existed. Also, in particular, Robert Morgan and Art Evans (formally of BBN) for excellence in the compilation, design and implementation of Praxis.

And finally, we thank the following individuals whose coding styles were analyzed statistically, and who provided the formal and informal material for this paper.

G. Aune	J. Krammen
D. Benzel	D. Kroepfl
R. Burris	D. Myers
A. DeGroot	R. Reed
D. McGuigan	J. Severyn
J. Greenwood	T. Sherman
J. Hill	J. Smart
W. Schaefer	K. Snyder
C. Humphreys	P. Van Arsdall
B. Johnson	J. Wilkerson

REFERENCES

Barnes, J. G. P. (1980). An Overview of Ada. Software-Practice and Experience, pp. 851-887.

Carlson, W. E. (1978). Ada: A Promising Beginning. Computer.

Evans, A., Jr., C. R. Morgan, J. R. Greenwood, M. C. Zarnstorff, G. J. Williams, E. A. Killian, J. H. Walker, and J. M. Duffy (1981). Praxis Language Reference Manual, University of California, UCRL-15331, Contract 3634909.

Evans, A., Jr. (1981). A Comparison of Programming Languages: ADA, PRAXIS, PASCAL, C, University of California, UCRL-15346.

Fisher, D. A. (1978). DOD's Common Programming Language Effort. Computer.

Hoare, C. A. R. (1981). The Emperor's Old Clothes. Communications of the ACM, pp. 75-83.

Ichbiah, J. D., J. C. Heliard, O. Roubine, J. G. P. Barnes, B. Krieg-brueckner, and B. A. Wichmann (1979). Rationale for the Design of the ADA Programming Language. ACM Sigplan Notices, Vol. 14, No. 6, Parts A and B.

Ledgard, H. F., and A. Singer (1982). Scaling Down Ada (Or Towards A Standard Ada Subset). Communications of the ACM, pp. 121-125.

Manes, K. R., et al. (1982). Novette, A Short Wavelength Laser-Target Interaction System. University of California, UCRL-88120, prepared for submission to the Sixth International Workshop on Laser Interaction & Related Plasma Phenomena.

Morgan, C. R. (1983). ADA: Its Motivation, History, Capabilities, Implementation. Intermetrics Inc. Lecture Notes.

Saib, S. H., and R. E. Fritz (1983). The ADA Programming Language: A Tutorial. IEEE Computer Society Press, IEEE Catalog EHO-202-2.

Simmons, W. W., et al. (1982). Nova Laser Fusion Facility: Design, Engineering, and Assembly Overview. University of California, UCRL-88700, prepared for submission to the Journal of Nuclear Technology/Fusion.

Suski, G. J., and F. W. Holloway (1979). Development of a Control-System Implementation Language (Praxis). Laser Program Annual Report, University of California, UCRL-50021, pp. 2-106

Suski, G. J., and F. W. Holloway (1979). The Evolution of a Large Laser Control System - from Shiva to Nova. IEEE Circuits and Systems, Vol. 1, No. 3 and Cover Photograph.

Suski, G. J., J. M. Duffy, D. G. Gritton, F. W. Holloway, J. E. Krammen, R. G. Ozarski, J. R. Severyn, and P. J. Van Arsdall (1982). The Nova Control System - Goals, Architecture, and System Design, University of California, UCRL-86827.

- -

TABLE 1

Ada Schedule MAY 1983

Formation of HOLWG	JAN	1975
Strawman	APR	1975
Woodenman	AUG	1975
Tinman	JUN	1976
Ironman	JAN	1977
Ironman Revised	JUL	1977
Competition Started	JUL	1977
Sandman	MAR	1978
Competition Narrowed	APR	1978
Steelman	JUN	1978
Pebbleman	JUN	1978
Language Chosen	APR	1979
Test & Evaluation of Design	OCT	1979
ROLM Ada Compiler	AUG	1982
NYU Ada Interpreter	NOV	1982
DOD Ada VAX Compiler	JAN	1983 (?)
INTEL Ada Compiler	JUL	1983
DOD Ada IBM Compiler	MAR	1984
Ada DBMS		1985
Ada Required in all New Embedded Systems		1990

TABLE 3

TABLE: USE OF FEATURES OF PRAXIS IN THE NOVA/NOVETTE CONTROL SYSTEM SOFTWARE

Culture	Compiler Writer	Prof.	Amateur	Novice	Total
# Modules	882	661	809	58	2410
# Lines Code	73062	109197	137043	5026	324328
# Comments	18845	26544	48296	1270	94955
# Blank Lines	16828	39359	54148	1816	112151
# Int. Semicolon	431	1046	1679	13	3169
# strings	5079	23015	27899	686	56679
# @	6700	1072	1150	18	8940
# ?	443	569	478	8	1498
# %	3584	2984	3006	63	9637
# :	33062	40042	39445	1214	113763
# :=	9173	11493	20426	596	41688
# *=	1587	609	565	15	2776
# Exports	4123	6574	4204	455	15356
Ave Line Length	22	24	30	25	34
Max Line Length	126	126	130	118	130
Identifier-uses	122712	174479	226538	6750	530479
Ave Identifers	5	7.3	8.7	8.4	7.4
Ave Leading Spac	3	6.7	6.5	7.1	5.8
--DIRECTIVES					
%INCLUDE	0	36	2	0	38
%DEFINE	61	109	201	3	374
%SET	118	295	333	17	763
%IF	1395	960	960	14	3329
%ORIF	2[illegible]	36	2	0	250
%OTHERWISE	127	23	25	0	175
%ENDIF	1395	961	960	14	3330
%IDENT	68	448	451	15	982
%SEGMENT	16	55	19	0	90
%ERROR	0	0	0	0	0
%MESSAGE	12	1	6	0	19
%PAGE	180	60	47	0	287
--MODULE INTERFA					
MAIN	537	136	169	24	866
MODULE	848	655	803	58	2364
ENDMODULE	845	653	802	58	2358
EXPORT	607	662	838	71	2178
IMPORT	2141	972	507	17	3637
USE	143	2262	3713	117	6235

TABLE: USE OF FEATURES OF PRAXIS IN THE NOVA/NOVETTE CONTROL SYSTEM SOFTWARE - Continued

Culture	Compiler Writer	Prof.	Amateur	Novice	Total
--MODULE SECTION					
PROCEDURE	1769	2336	2081	61	6247
ENDPROCEDURE	1667	2178	2019	61	5425
PARAM	411	699	638	7	1755
ENDPARAM	411	699	638	7	1755
FUNCTION	610	507	395	33	1545
ENDFUNCTION	584	496	385	33	1498
RETURNS	610	507	389	33	1539
FORTRAN	4	10	118	1	133
FORWARD	55	121	32	0	208
CODE	11	30	26	1	68
ENDCODE	11	30	26	1	68
INTERRUPT	11	22	83	0	116
-- COLLECTIONS					
DECLARE	2028	1474	2152	92	5746
ENDDECLARE	1134	1216	1649	72	4071
STRUCTURE	250	759	326	28	1363
ENDSTRUCTURE	245	759	325	28	1357
ARRAY	1217	2203	2820	87	6327
-- ATTRIBUTES					
ADDR_OF	1	72	158	0	231
ALIGNED	0	7	4	0	11
FILL	6	362	19	9	396
HIDDEN	8	0	0	0	8
IN	846	2610	2964	33	6453
INOUT	543	902	682	6	2133
OUT	82	374	1209	3	1668
OPTIONAL	994	1154	243	0	2391
IS	1438	1585	1470	167	4660
OF	1269	2203	2826	87	6385
INITIALLY	3464	2510	2407	82	8463
LOCATION	16	42	130	79	267
PACKED	687	1419	1248	80	3434
RANGE	74	31	51	2	158
READONLY	1	0	0	0	1
REGISTER	2	0	0	0	2
REF	1052	2644	3988	24	7708
SET	52	0	5	0	57
TABLE	166	107	484	8	765
VAL	396	1155	625	10	2186
VOLATILE	17	116	107	32	272
SEGMENT	13	116	102	0	231
STATIC	1128	1130	3009	95	5362
FALSE	1043	1459	3365	4	5951
TRUE	1029	1314	2383	61	4787
NIL	1118	227	4	0	1349
UNLOCKED	34	45	0	0	79

TABLE: USE OF FEATURES OF PRAXIS IN THE NOVA/NOVETTE CONTROL SYSTEM SOFTWARE - Continued

Culture	Compiler Writer	Prof.	Amateur	Novice	Total
-- TYPES					
BIT	303	4846	3500	205	8854
BOOLEAN	752	1111	1087	19	2969
CHAR	1194	1006	1328	100	3628
INTEGER	2633	7555	6539	243	16970
GENERAL	66	250	12	0	328
POINTER	250	225	29	81	585
CARDINAL	223	125	164	32	544
LOGICAL	358	2775	479	142	3754
REAL	503	477	1301	13	2294
-- OPERATORS					
ABS	78	92	65	0	235
AND	1214	624	955	60	2853
XOR	2	0	3	0	5
MAX	34	560	11	0	605
MIN	113	75	55	0	243
NOT	430	366	794	33	1623
OR	1001	674	640	13	2328
FORCE	277	858	722	78	1935
ROUND	17	0	0	0	17
CEILING	8	0	0	0	8
FLOOR	17	0	1	0	18
HIGH	45	106	49	1	201
LOW	13	93	45	0	151
PRED	48	1	0	0	49
SUCC	77	1	7	0	85
SIZEOF	5	0	0	0	5
SWAP	19	0	0	0	19
-- HEAP					
ALLOCATE	229	44	0	0	273
FREE	39	71	2	0	112
-- EXCEPTIONS					
ARM	28	0	41	0	69
DISARM	2	0	0	0	2
CATCH	368	101	119	0	588
EXCEPTION	102	89	16	0	207
FAILED	0	0	1	0	1
GUARD	368	101	119	0	588
ENDGUARD	368	101	119	0	588
RAISE	292	370	35	0	697
RERAISE	5	6	18	0	29
FINISHING	14	0	0	0	14
ASSERT	144	77	87	3	311

TABLE: USE OF FEATURES OF PRAXIS IN THE NOVA/NOVETTE CONTROL SYSTEM SOFTWARE - Continued

Culture	Compiler Writer	Prof.	Amateur	Novice	Total
-- REGIONS					
REGION	63	27	0	0	90
ENDREGION	63	27	0	0	90
INTERLOCK	37	16	70	0	123
RETRY	0	8	0	0	8
-- FLOW CONTROL					
BLOCK	34	1	33	0	68
ENDBLOCK	34	1	33	0	68
BREAK	146	85	216	30	477
FOR	979	1540	2050	88	4657
ENDFOR	972	1538	2036	88	4634
TO	747	1191	1903	73	3914
DOWNTO	47	44	8	0	99
THEN	278	192	101	8	579
IF	3812	3898	5062	94	12866
ORIF	363	658	1049	85	2155
OTHERWISE	1023	1201	1295	27	3546
ENDIF	3799	3898	5063	94	12854
DO	5591	6322	8341	300	20554
LOOP	77	57	334	2	470
REPEAT	148	154	242	86	630
UNTIL	148	154	244	85	631
RETURN	468	430	747	27	1672
SELECT	391	314	992	9	1706
FROM	2532	1286	1499	26	5343
ENDSELECT	390	313	991	9	1703
CASE	2146	1951	4495	68	8660
DEFAULT	409	225	537	4	1175
WHEN	126	138	98	4	366
ELSE	126	138	98	4	366
WHILE	526	226	168	36	956
ENDWHILE	376	172	163	32	743
WITH	6	120	1	8	135
ENDWITH	5	120	1	8	134
UPON	10	11	40	0	61
ENDUPON	10	11	40	0	61
LEAVE	10	11	40	0	61
VIA	28	100	246	0	374
THROUGH	10	11	40	0	61

```
//
//      Example of a Complex Number Package in Praxis
//

module COMPLEX_LIB

    use MATH_LIB

    export  COMPLEX, COMPLEX_SUM,
            COMPLEX_PRODUCT, MAGNITUDE

    declare
        COMPLEX is structure
            REAL_PART               : real
            IMAGINARY_PART          : real
        endstructure
    enddeclare

    function COMPLEX_SUM ( X, Y: in ref COMPLEX )
        returns SUM : COMPLEX
        SUM.REAL_PART       := X.REAL_PART + Y.REAL_PART
        SUM.IMAGINARY_PART  := X.IMAGINARY_PART + Y.IMAGINARY_PART
    endfunction {COMPLEX_SUM}

    function MAGNITUDE ( X : in val COMPLEX )
        returns R : real initially
        FSQRT ( X.REAL_PART**2 + X.IMAGINARY_PART**2 )
    endfunction {MAGNITUDE}

endmodule {COMPLEX_LIB}

//
// use of above
//

Main Module TEST

    use COMPLEX_LIB

    declare
        X       : COMPLEX
        Y       = COMPLEX ( REAL_PART: 1.0,
                            IMAGINARY_PART: 2.0 )
        Z       : COMPLEX initially Y
        R       : real
    enddeclare

    X := COMPLEX_SUM ( Y, Z )
    R := MAGNITUDE ( X )

endmodule;
```

```
--
--      Example of a Complex Number Package in Ada
--

package COMPLEX_LIB is

    type COMPLEX is record
        REAL_PART       : real;
        IMAGINARY_PART  : real;
    end record;

    function "+" ( X, Y: COMPLEX ) return COMPLEX ;
    function ABS ( X    : COMPLEX ) return COMPLEX ;

end COMPLEX_LIB;

--
-- And NOW the package body, that is, the implementation
-- of the specification.
--

with MATH_LIB;  -- a package containing FSQRT function

package body COMPLEX_LIB is

begin

    function "+" ( X, Y: COMPLEX ) return COMPLEX is
    -- an overloaded operator
    begin
        return ( X.REAL_PART + Y.REAL_PART,
                 X.IMAGINARY_PART + Y.IMAGINARY_PART);
    end "+";

    function MAGNITUDE ( X    : COMPLEX ) return real is
    begin
        return FSQRT( X.REAL**2 + X.IMAGINARY**2 );
    end MAGNITUDE;

end COMPLEX_LIB;

--
-- use of same in a program fragment
--
    declare
        use COMPLEX_LIB;
        X       : COMPLEX;
        Y       : constant COMPLEX := ( 1.0, 2.0 );
        Z       : COMPLEX := Y;
        R       : REAL;
    begin

        X := Y + Z; -- use complex addition
        R := MAGNITUDE(X);

    end;
```

Fig. 1. Comparative Example of Praxis and Ada

Table 2

	Culture				
	Compiler Writer	Professional	Amateur	Novice	Total
Number of individuals	4	8	9	7	28
Lines of actual code	73,062	109,197	137,043	5,026	324,328
Percentage of lines of comment per line of code	26%	24%	35%	25%	-
Assignments	9,173	11,493	20,426	596	41,688
Number of separate modules	882	661	809	58	2,410
Number of exported items	4,123	6,574	4,204	455	15,356
No. of identifier references	122,712	174,479	226,538	6,750	530,479

DISCUSSION

MacDonald: I would like to hear a comment by Dr Suski regarding the usefulness or otherwise of PASCAL as a stepping stone for moving towards ADA.

Suski: PASCAL is a good stepping stone, and in many implementations of PASCAL, particularly the ones with separate compilation features, it can prepare you for changing to ADA. ADA is still perhaps not complete in terms of a language for real-time systems and it really needs to be recognized in that context.

MacDonald: I am really thinking of a person who is finding FORTRAN inadequate, but who can't wait for ADA to arrive.

Suski: My advice is to use PASCAL, but you should have a very close look at the particular implementation.

Sloman: I would like Dr Suski to comment on the use of ADA for distributed systems as opposed to merely imbedded ones. In our research we looked at ADA and we thought that it would not be an appropriate language to use.

Suski: Our application is highly distributed and the language which we used, PRAXIS, is, in my opinion, similar enough to ADA for us to make a meaningful comment. We feel that we can state that PRAXIS worked very well in our system and helped us to develop programs, in different computers, which communicated to each other correctly because their data structures were essentially the same. Also, we have been able to deliver a system on time!

Sloman: You mentioned however that you do not have anything like the Rendezvous of ADA and in my opinion this construct in ADA will present problems in a distributed system.

Suski: PRAXIS certainly uses a different technique, but in my opinion the Rendezvous, while it can be used in distributed processes, certainly looks to me as if it was initially developed to be used for co-resident routines in the same processor - or at least processors sharing the same memory.

Kopetz: When I set about designing our system, we looked very closely at the Rendezvous construct and we found it completely inappropriate for real-time distributed systems.

Imamichi: It was claimed by Dr Suski that ADA is a most useful language in the development of a distributed system. I would like him to outline his prime reasons for this conclusion.

Suski: The statement was based primarily from the standpoint of management of software development. In ADA the strong typing, the separate compilation mechanisms, etc. but not necessarily the task synchronization techniques, make it an ideal language. You could obviously use any language within a distributed computer control environment, but I think ADA is a suitable language. It may not have everything you want, but it is better than many of the languages which have been used up to now. Also the point that it will receive long term support clearly leads towards ease of long term maintenance.

Plessmann: Both ADA and PRAXIS have very strong typing. In my opinion it only makes sense to have this feature if you have a run-time check. What would happen if all the features are checked during run-time? Would this derate the normal performance by 10% or 20%? If this happened what would be the effect if the computer performance already is at 80% of capacity?

Suski: You have to look at this issue from a number of standpoints. Firstly, because of the way the language is supplied within a package, the type-checking can be done at compile time to a large extent. Where the problem comes in is when you start to enter procedures in separate packages, or procedures which are called from any external source which you cannot trust. As I mentioned, I believe that this is one of the great concerns about ADA, i.e. that run-time checking will impact its efficiency. There are currently a large number of people who are dealing with the issue in great detail. For instance, they are looking at ways of evaluating values of parameters as they come in and, based on these values, eliminate many of the checks on the main parameters - just because these parameters will not be used!

Sloman: I disagree with this approach. As I mentioned, even for the communication between separately compiled modules, we only do static type-checking. We do not believe in any form of run-time type-checking. This is because the only way things can go wrong is by corruption and this can be catered for by normal error detection mechanisms.

Suski: Unfortunately the definition of ADA as it stands right now includes the formality of run-time type-checking. In PRAXIS we have the ability to turn that feature off! ADA unfortunately does not allow it.

Heher: In our area we have been using a language for some time which has been in existence for over ten years and that is RTL/2. This does strong type-checking, either at compile time or at the time when you are linking or task building the system. Again you do not need any run-time checks. We also found that a high-level language of this sort, which has many of the features of PRAXIS, is very effective. A penalty of not more than about 5% compared to assembler code is typically incurred.

Maxwell: What do you think the impact of ADA and PRAXIS will be when the super-fast computers emerge at a price which makes them viable in process control?

Suski: Clearly things will get much better! Right now there is a question as to whether ADA, with the current generations of machines, is going to do all we need in all cases. But given another generation of machines which are more powerful and faster, things could certainly get much better.

Atkins: Will the advent of ADA, in fact, dramatically influence future machine architectures?

Suski: My feelings are that ADA as it stands right now is not going to affect machine architectures. I believe that the work which is being undertaken in architecture may well influence ADA. ADA is really built around the architectures as they currently exist. New architectures which are starting to emerge might well require ADA in order to take full advantage of them.

Heher: This question, of course, worries us a lot. We are talking about extensions to ADA and things that have been found missing from ADA. This seems to imply that ADA is not going to be a fixed language for some time.

Suski: No - there is too much pressure to keep it as it is for a while. If ADA is going to change it is going to be for extremely good reasons, and it will be in response to a lot of common pressure, but this will not occur in the very near future.

Harrison: In fact when the DOD started the effort, they made the statement several times that once they determined the final specification, it will remain unchanged for 10 years. This caused a certain amount of consternation in the Standards Activity, which typically requires a 5-year review. However, the designers of ADA basically said that, even though we might make a mistake, we will want the language to remain the same unless it is a fatal mistake, because of the advantages of a stable language.

LaLive d'Epinay: The possible disadvantages of not allowing subsets have been mentioned. Don't we confuse two things - what is allowed in the subset of a compiler and what is allowed in a subset of the range of the language which is used for a specific project? You are totally free to limit the size of the language you are using so why bother to have a compiler which can compile everything possible within the official definition of the language?

Suski: In general I agree with you, and in my paper I pointed out that I would like to see some management-directives directed at some of the implementations. This would allow us to control subsets used in a particular project. Of course, part of the pressure on subsets comes from the people that do not think ADA can be efficient enough in the form in which it is right now, in some machines. Thus, they are looking for subsets so as to allow them to use ADA more practically on smaller machines. I believe that pressure is going to dissipate as machines become more powerful.

DATA CONSISTENCY IN SENSOR-BASED DISTRIBUTED COMPUTER CONTROL SYSTEMS

I. M. MacLeod

Department of Electrical Engineering, University of the Witwatersrand, Johannesburg, South Africa

Abstract. The possibility of holding the data collected from the plant in a functionally distributed data base is one of the interesting possibilities offered by distributed computer control systems. In the paper, the concept of state information is clarified and a new definition is given for the validity and consistency of state values in real-time sensor-based systems. A real-time distributed algorithm for establishing a valid and consistent distributed representation of the data collected from sensors by periodic sampling is presented.

Keywords. Distributed computer control systems, real-time systems, real-time programming, process control, distributed data base, distributed algorithms, data consistency, periodic sampling.

INTRODUCTION

Proper handling and representation of the data collected from the plant is one of the most important aspects of any real-time computer control scheme. This data forms the foundation for all higher control and information processing functions. The possibility of holding such data in a functionally distributed data base is an important advantage of distributed computer control systems (DCCS), for reliability and modularity reasons. There are, however, many problems in establishing a distributed representation of the plant data and ensuring its consistency.

The emphasis in this paper is on developing methods that allow such real-time sensor-based DCCS to be constructed largely from standard components such as single-board computers, communicating via a multi-access packet broadcast communications system such as a local area network. We assume that the processors in the system have a high degree of autonomy and are not tightly synchronized. Systems of this type offer great flexibility and potentially support the addition of redundant processors in a way that is transparent to software.

In this type of system co-ordination can be achieved by arranging for each processor to have access to a global physical time and the work in this paper shows how the problems associated with establishing a valid and consistent distributed real-time data base can be solved using this approach.

STATE INFORMATION AND SAMPLING

A control scheme comprises two subsystems; the physical plant system and the control system. From the point of view of the control system, the physical plant system is its environment.

The concept of state as it is used in this paper relates to the condition of those physical objects in the environment of the control system whose behaviour can change with time and to which stimuli can be applied and the responses observed. State information can be obtained in the form of signals from sensors that measure attributes that are characteristic of objects in the environment.

It is necessary to have a real-time representation of the state of the environment in the control system and because of the sequential and discrete-time nature of the digital computer, this must be done on the basis of repetitively sampling the continuous-time signals at defined points in time and assigning the resulting numbers to variables. The values resulting from this sampling process are assumed to adequately represent the properties of the environment during the interval from the sampling time until the time the next sample is due to be taken.

An individual state value resulting from a sampling operation is thus invalidated by the passage of time in a real-time system and two time attributes must be associated with each state value to define it unambiguously: the time at which the sampling operation was performed and the time until which the value is valid. While all centralized real-time systems implicitly recognize the above factors, in the distributed real-time systems under consideration in this paper, it becomes much more important to consider them explicitly. The reason for this is the additional complexity introduced by unreliable communication via messages and the loose synchronization of processes.

The foregoing discussion has concentrated on

objects in the environment of the control system. Objects internal to the control system itself also have states associated with them and the following definition therefore applies generally:

Definition (State Information)

State information in a real-time system represents attribute values of objects and is valid for a known time interval.

The control system variables that hold state information obtained from sensors collectively define an image of the environment and we will call them Image Variables in order to distinguish them from internal state variables. The collection of image variables, which may be spatially distributed throughout the control system, constitutes a distributed real-time data base, which we refer to as the Image Set.

The inherent granularity of the operations of sampling and updating the image set will be carefully considered in the following analysis of the properties of the image set.

THE IMAGE SET

Real-time control systems are usually organized as a number of tasks and these tasks are executed cyclically and in parallel. For example, an alarm reporting task periodically scans image variables and checks if prescribed limits have been exceeded, or a control algorithm periodically reads image variables and computes any necessary adjustments. The organization of such systems is greatly simplified if facilities are provided in the system that allow application tasks to gain access to that image set without requiring tight synchronization with the cycles of the sampling tasks or with one another.

Construction of such a valid and consistent image set is a difficult problem. Consider the simple situation illustrated in Fig. 1, where successive sampling instants of two processes (possibly running in separate processors) that are both sampling the same switch are shown:

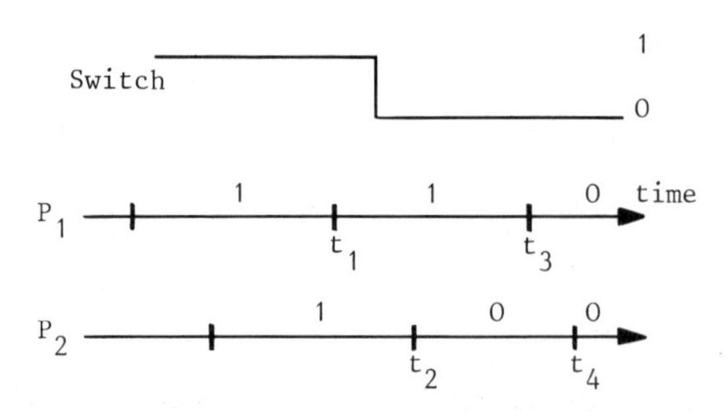

Fig. 1. Two Processes Sampling a Switch

During the interval (t_2, t_3) the values of the local image variables updated by the two processes P_1 and P_2 are in conflict. In some distributed applications it might be necessary to send the results of the sampling operations of P_1 and P_2 as messages to a third process, P_3 say, and for P_3 to have access to a local subset of the image set that contains, as one of the image variables, the state of the switch. There is obviously a consistency problem here, as the values in the two messages are different. In this case, the unknown message transmission delays and the possibility that messages can get lost must also be taken into account.

At first sight, closer synchronization of the sampling instants t_1 and t_2 may appear to be a solution to this consistency problem, but this is not the case. Firstly, two processes in a distributed system can never be exactly synchronized and secondly, even if exact synchronization were possible, slight differences in the logic thresholds of the input circuits would mean that one process might just see the new state of the switch while the other would just miss it.

In the case where two or more switches are being sampled by different processes, similar reasoning to the above shows that it is possible for combinations of states of the switches that never existed physically to be generated in the control system. This could have most serious consequences in a critical control scheme and it is therefore important to find protocols for the construction of, and access to, a valid and consistent image set.

It is interesting to note that Lamport (1978) has identified a related problem in connection with a nationwide system of interconnected computers. He refers to the situation where there is uncertainty as to whether one environment event preceded another as "anomolous behaviour" and shows that a practical solution must be based on allowing the processes to access physical time.

DISTRIBUTED SYSTEM MODEL

In order to facilitate discussion in the following sections, we develop in this section a simple model for real-time distributed computer control systems. This model applies to a set of distributed processors interconnected in a high-speed local area network environment as described by MacLeod and Rodd (1982).

Each processor can support a number of processes. Each process in the system, P_i, has a set of local variables X_i. There are no shared variables and processes may only interact by sending messages. Messages may be sent to individual specified destinations or broadcast to a group of destinations. Messages are subject to an unknown and variable transmission delay and may get lost. Each process has access to the system global physical real-time (the significance of this assumption and implementation issues are discussed in MacLeod (1983)).

For the transmission of the state values defined previously, it is adequate to model the handling of messages simply as the assignment of the value contained in a message to a local variable belonging to the remote destination process, after an unknown delay. Such a local variable is read-only to processes at the destination and an arriving state value simply overwrites the previous value stored in the variable, as for normal assignment.

A possibility frequently referred to in the next section in connection with transmission of the result of a sampling operation is that the validity time can be sent together with the value in a message. This allows the receiver to know when the image variable created from the value in the message ceases to be valid (defined in the next section). We call processes that have access to sensors and periodically transmit the sampled values as messages, observer processes and the receiver processes that use such messages to reconstruct the necessary valid and consistent image subsets, user processes.

Finally, note that only some of the local variables of a process are used to hold the image subset V_i, i.e. we have $V_i \subset X_i$. The values that may be held in each local image variable can either be of type Real or type Boolean, i.e. the jth local image variable of process i, X_{ij}, if it is of type Real can hold values x_{ij}, where $x_{ij} \in \mathcal{R}$, or if it is of type Boolean, can hold values x_{ij}, where $x_{ij} \in \{true, false\}$.

The Real image variables are used to hold the values obtained from analog measurements. The Boolean image variables are used to provide logical descriptions of the attributes of objects in the environment, for example TEMPERATURE > 100.0°C or the state of a switch.

In the following sections we concentrate exclusively on the Boolean image variables because consistency problems are more apparent and more serious for these. The results of the analysis can easily be extended to treat the Real image variables.

DEFINITION OF VALIDITY AND CONSISTENCY

As a first step towards developing algorithms for establishing a useful image set, we must clarify the meaning of validity and consistency as these terms relate to image variables.

Consider as an example two processes P_1 and P_2. P_1 has access to a sensor and produces a Boolean value describing the state of some object in the environment of the system (P_1 is an observer process). P_2 needs access to this same state as an element of its image subset. Process P_1 must therefore send the value in a message to P_2 every time it performs a scan of the sensor. We need to know the real-time inverval during which it can be guaranteed that the sample value stored in the Boolean image variable local to P_2 adequately represents the state of the environment. We say that the value is valid when this is so.

Definition (Valid)

> The value stored in an image variable is valid in the time interval during which it is equal to the result of a sampling operation performed not more than t_v seconds previously.

The validity interval parameter t_v must be chosen for a given implementation, as must the time between samples t_s. Depending on t_v and t_s and the delay experienced by the particular message carrying a value to a user process, the following cases can arise.

A. There can be time intervals during which the value stored in an image variable at the user process is no longer valid and a new value has not yet arrived. In this case we introduce a special value ω - the undefined value - as an indicator to any user of the image variable that the contents are not to be used.

B. There can be an overlap of the valid intervals of successive samples.

In the remainder of this paper, we consider an implementation in which the validity interval given by t_v is equal to t_s, the time at which the next sampling operation is due to be performed. This means that no new state values are transmitted by an observer process before the validity of previous values has expired at the remote user processes. We therefore restrict discussion to case A above and we must assume that under normal circumstances the message transmission delay is less than the time between samples. The situation is illustrated in Fig. 2 which shows two out of an on-going sequence of sampling instants and the corresponding assignments after message transmission.

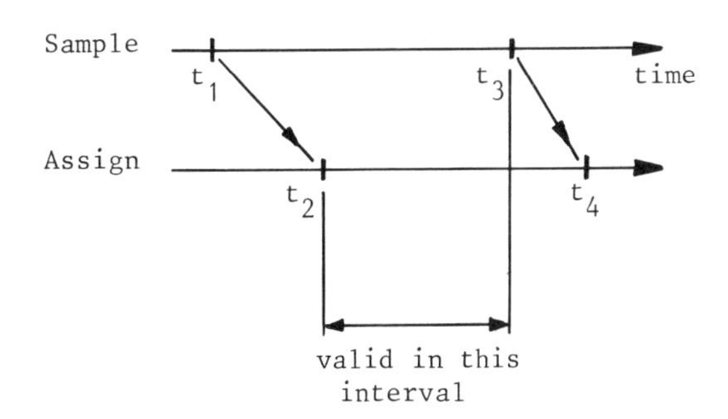

Fig. 2. The Validity Interval for a Sampled Variable

Note that in this example we use t_3 and not t_4 to terminate the valid interval because due to the variable message transit delay it is not known at the sender when t_4 occurs. It may in fact occur at different times for different receivers in a broadcast situation. If the message is lost, it may not occur at all. The content of the image variable is not valid in the interval (t_3, t_4) and the implementation at the receiver must ensure that the undefined value is loaded into the variable at time t_3.

Consider now one particular image subset V_i. In general, a happening (event) occurring in the environment may or may not affect the plant states reflected in V_i. Call an event that does cause a change in one or more of the plant states in V_i a significant event for V_i. In general, a significant event will be asynchronous with the sampling of the observer processes that are supplying values for V_i. As with the example in Fig. 1, there is a danger that a state value resulting from a sampling operation performed just before the event is valid concurrently with one from just after the event, with the consequence that combinations of states that never existed physically can appear in the image subset. This is a consistency problem and leads to the following definition:

> Definition (Consistent)
>
> An image subset is consistent in a time interval if no significant events occurred in the time interval between the earliest and most recent sampling times corresponding to its valid values.

Now, from the foregoing discussion it is clear that there will be times when the values stored in an image subset are all valid but not consistent. In order to simplify the rules for accessing an image subset, we introduce the following additional condition that must be met by an implementation:

C1. At any time and for any sequence of events that can affect the state of the environment, if all the image variables in any image subset are valid, then the implementation must ensure that they are also consistent.

If C1 is met, once a user process has accessed a subset of valid values(as an atomic operation), it can assume that the values are consistent. A process could be forced to wait until all the values in a subset are valid.

A DISTRIBUTED SAMPLING ALGORITHM

One way of implementing a system that satisfies consistency condition C1 is for every observer process to control the interval during which each state value that it transmits is valid. The responsibility for ensuring consistency is thus with the observer processes and in this approach, when all the values in any image subset are valid, the user process hosting that subset can safely assume that the subset is consistent. No special action other than the mechanism for setting a value undefined after its validity time has expired is needed at the receivers. This mechanism is in any case necessary to handle lost messages.

From C1, the following implementation rules which must be obeyed on an individual basis for each image variable can be stated:

IR 1. Once an observer process has detected a change of state, it must wait long enough before transmitting a new value for all image variables that could also change as a result of the same event (irrespective of their source observer processes) to take on the undefined value.

According to the earlier message handling model, this can be implemented by each observer process simply omitting to send a message corresponding to the first sample (and possibly subsequent samples) after detecting a changed state. The validity time interval of the previous value held in the affected image variable is then automatically terminated at each remote user site. The effect of this is that at each user process there is a "shrinking phase" during which the number of affected image variables remaining valid reduces to zero. Valid values from samples taken before the event was detected are thus all removed before any new values are transmitted, thereby preventing conflict.

IR 2. Before a new valid value can be transmitted, each observer process must have performed at least one sampling operation since the event.

This ensures that all observers have detected the change.

IR 3. An image subset may only be read when all its elements are valid.

The main practical difficulty concerns IR 1 and IR 2. These imply that each observer process must know about the actions of all the other observer processes. This could be achieved by interchanging additional messages, but this is not considered to be practical. An alternative approach based on the real-time properties of the sampling operation is pursued in the remainder of this section.

It is traditional in digital control systems to use periodic sampling of the signals obtained from sensors. In the ideal case, the behaviour of the sampling process is modelled by the following sampling rule:

SR 1. $t_{n+1} = t_n + \Delta$

Here t_n is the point in time at which the nth sampling operation is performed and Δ is the sampling period. Note that no synchronization of the sampling instants of different observer processes is implied. It is almost impossible in practice to implement SR 1 exactly, due to operating system overhead and software non-determinacy, but many systems approximate it quite closely.

A weaker "average periodicity" that can be guaranteed by virtually any implementation of sampling is:

SR 2. $(n-1)\Delta + t_o < t_n < n\Delta + t_o$

This states that a sampling operation is performed once and only once during each time interval of duration Δ.

Returning now to the distributed sampling algorithm, if it is known that all the observer processes contributing to an image subset obey SR 1, then any individual observer process P_i can be sure that in the time interval between its nth sample and (n + 1) th sample, every other observer process will perform exactly one sampling operation. Implementation of the sampling algorithm is now straight-forward:

All observer processes perform periodic sampling according to SR 1 using the same sampling period, based on the global real-time. If an observer process detects a change of state after sample n it must not transmit the new value. The new value can be transmitted after sample n + 1 if there has been no further change of state. User processes may only read an image subset if all the values in the subset are valid, based on the validity time supplied by the observer processes and the global real-time.

In the case where all the observer processes obey SR 2, then between the nth and (n + 1) th sample of any observer process P_i, it is only known that each other observer process will perform either 0, 1 or 2 sampling operations. The possibility that zero sampling operations may be performed violates IR 1 and IR 2. However, between sample number n and sample number (n + 2) each other process will perform either 1, 2 or 3 sampling operations. This means that the above sampling algorithm can again be used, except that an observer may now not transmit a new value until after sample (n + 2), if the change of state was detected after sample n.

Detailed proofs of the correctness of the sampling algorithm described above are available elsewhere (MacLeod, 1983). The more complete development shows that the algorithm also works for the case of redundant observer processes and lost messages. It is thus possible to use this approach to establish a valid and consistent image set with high reliability in a distributed system. The assumption concerning a global real-time can be relaxed to allow a set of local clocks that are synchronized to within some known accuracy.

CONCLUSION

An approach for establishing a functionally distributed real-time data base in distributed computer control systems has been presented. The distributed algorithm described makes explicit use of a global physical time to solve the consistency problems that can arise. The algorithm is easy to implement if processes have access to synchronized real-time clocks and also supports redundant measurement sampling.

ACKNOWLEDGEMENTS

Many valuable discussions with Professor H Kopetz of the Technical University of Vienna have greatly enhanced this work. The guidance of Professor M G Rodd of the University of the Witwatersrand is gratefully acknowledged. The author also wishes to thank the University of the Witwatersrand for assistance in the production of this paper. This work would not have been possible without the support of AECI Limited.

REFERENCES

Lamport, L. (1978). Time, clocks and the ordering of events in a distributed system, Comm. A.C.M., 21, 558-565.

MacLeod, I.M. and Rodd, M.G. (1983). Interprocess communication primitives for distributed process control, in Software for Computer Control 1983, Ferrate and Puente eds., Pergamon Press, Oxford.

MacLeod, I.M. (1983). A study of real-time issues in distributed computer control systems, PhD. thesis, University of the Witwatersrand, Johannesburg.

DISCUSSION

Gellie: I can understand that within a single component it is important to know whether two observers have sampled. This is clearly necessary to check correctness and consistency between two readings but I do not see why at any receiver end, it is important to know that two observers have knowledge as to whether the other observer has, in fact, sampled.

MacLeod: What I am saying is that any given observer should be able to know that others have sampled, and are obeying periodic sampling. In the case of strict periodic sampling one can ensure consistency after only one scan. In the case of weaker types of periodicity two scans would be required. This implies that one observer will know without interchanging any other information that a consistent state has been obtained.

Gellie: My point is that as far as an observer is concerned it is always sending what it knows to be correct and valid data unless it is not working. If it suddenly discovers that, after it has done its sampling, there has been a state change then it recognizes that the data which it sent out, which it thought at the time was correct and valid, is no longer consistent with the environment. It should then simply refuse to send data until the next time.

MacLeod: The point is that the observer must know what all others are doing as well, and it must be sure that every other observer has also taken a sample since the event occured, otherwise there is a danger of inconsistency.

Gellie: It would seem to me then that you are considering the problem of fault checking back at the sender end as well as recognition of consistency and validity at the receiver end.

MacLeod: This is correct. We are actually looking at both ends as a combined operation.

Kopetz: If you do not make the assumption that Mr MacLeod is making, then you can get inconsistency at the receiver, because it could be that one node goes to a new value and the other one has not changed its old value yet; you are therefore inconsistent.

Sloman: Does Mr MacLeod only allow observed values in his image data set or does he allow derived values as well?

MacLeod: We certainly allow derived values as well. A typical example would be say, a logical variable indicating temperature greater than 100 degrees centigrade - which is derived from plant data. We clearly must allow both types.

LaLive d'Epinay: I would like to formulate a provocative question. Is it not true that you are starting at the wrong end when you are discussing consistency of data if you only look at the special cases of digital information. In the plants that we are involved with, and in most practical cases, we have analog information and according to Mr MacLeod's definition, any analog information would never be correct in his system. So should we not try a totally different approach to find out whether the system is consistent? For instance, in a power distribution system a switch has about ten different states - it begins with closed states which are then ordered to be opened. But you will have the situation when some are opened and others are closed but this cannot be confirmed. This is the way we try to differentiate then, for instance, the case where one sensor stays closed and the other stays open. You know that this can occur so you test it with some form of timeout. My feeling is that Mr Macleod has only looked at a special case and should go another way to find the general solution - like the state estimation in a power distribution system for example.

MacLeod: I can explain my approach to analog values and that is that I have concentrated in my work specifically on the status case because it is well formulated and the problem is more critical here. In a sense if you think of an analog value you could apply the same reasoning to every bit in that value. In practice the consistency problem is not so severe with analog values and we will never get two transducers to provide exactly the same binary pattern anyway. I think you must accept slight variations in analog values, but the feeling is that the same model basically applies and you could use the same algorithms.

Kopetz: We are clearly back to the question of definition. What do you actually mean by consistency, validity and correctness? There are clearly some differences in these definitions. Mr MacLeod has produced one form of definition and this appears reasonable. In essence we feel that there are two ways of approaching the problem. You can either do as Mr MacLeod has done and require total synchronism with the global time of the plant, or you could allow a certain delay between the state of the system and the state of the plant. This is the way, in our work, we would rather go. We define consistency which means that you bound the time to your image set and, say, within the last units of time this set has actually occurred in the plant, and was a correct representation of the plant - within the limitation. There are good reasons for looking at the problem in both ways, but only time will show which of the concepts and definitions is workable in the more general case. I would like to make the point that this is an area in which things are still not quite clear.

Van Selm: Extending Mr MacLeod's ideas, do you think that you could extract the concepts away from the user end towards the plant end? If you could synchronize the samplings so that you never have any discrepancy in terms of the system observers you clearly would appear to simplify the matter considerably. You could then get a perfect observer, or several perfect observers, in processes that were synchronized to the global time and work from this basis.

MacLeod: Our feeling is that it is never possible to exactly and precisely synchronize two independent processors so that there could always be an interval during which one is just sampled and one has not yet sampled. Even if you could synchronize them exactly, different logic thresholds in the electronics would still cause the same problem. One processor might see the change and one might just not.

Rubin: I would like to say that Mr MacLeod's model seems to me to be rather simplified in that it does not allow for incorrect measurements. In other words, measurements which do not indicate the value of a state of a variable, but only approximate to it. If one used the Kalman approach, one is inclined to use assessments of the states, plus all new measurements plus values assigned to each of these, to determine the present state. I would also like to question the terminology of consistency and correctness as it seems to me a very rigid one which I would be inclined to see as being useful if you were going to make a tactical decision rather than a control adjustment.

Kopetz: I agree with the comments about the preciseness of these terms. To this point, as far as I know, there has been no attempt to produce definitions in the way I have. So my work is really an attempt to at least have a starting point. Clearly we are moving into the topic of errors which brings a completly new dimension into the problem domain. This is the problem relating to making an observation error, or a measurement which fails. In general we are still struggling with the case in which we do not consider any faults at all. Clearly Dr Rubin's comment is valid, and this dimension, which includes the error handling, still has to be looked at in great detail. Currently we have to get the first problem sorted out and this is the case in which we assume that the observer sees everything actually which happens and there are no faults in the observation. We look at this in the time domain. Secondly I don't see the difference between a tactical decision and a control adjustment. My feeling is that we should define an architecture which is clean from the very beginning and in which both cases are coped with. We might be willing to allow a little bit of variations in the system but essentially we feel that if you design a distributed system architecture there should be sound reasoning behind it and we should not make a difference between the two problems.

Rubin: My feeling is still that the system designer would treat the computer inconsistencies in different ways when deciding how to operate in the two cases.

Kopetz: But if you have a real-time data base in your system and this is the basis for all the decisions you make, and you do not know how good it is, you do not know what kind of inconsistenties and deviations are in it. From an architectural point-of-view then, I feel you are not in a good position to make any kind of reasonable decision. So I would rather go for the approach in which you attempt to ensure a consistent representation in the environment of your plant at all times.

Heher: I would like to follow on the comments made by Dr Rubin. From an observation in the chemical industry, you clearly have more inputs than you have outputs. It really comes down to the question of tactical decisions in which you might read many variables but you take relatively few actions. In some cases there are perhaps 1000 inputs and only six actions being produced, therefore is it not much simpler to try and take these decisions at the output?

In other words, re-doing the calculation again a second time and ensuring that you get consistent answers. It is, in fact, almost the same as the sampling delay in which you are waiting until the data is consistent before you make a decision. But, because there are so few decisions that have to be taken, maybe this is a more simple approach?

Kopetz: I feel that if you do have a corrupted real-time data base in your system, you are in no better position to take a major decision than a minor decision. I do not think you will be able to draw the line between these two. Furthermore if you do not solve the problem cleanly, right at the beginning without the fault space considered, you will have a completely new dimension coming into the problem.

Gellie: What concerns me are critical applications such as an aeroplane's control system. If we have sensors which are looking at the temperature of the engine and, say, indicating that it is about to blow up then we meet up with the obvious problem that the system says that we need to have a couple of looks at the sensors before we are really sure that a dangerous situation has occurred. This could then mean that we do not have sufficient time to eject before the engine explodes! So is the answer then to put in another sensor into the system to ensure that sensor failure will not give a false indication. However this would only mean that we have to wait even longer the next time round!

MacLeod: This is clearly a very important point. We are talking about active redundancy in this approach. If we have the possibility that the actual values used may come from more than one sensor and that they may differ because of physical faults in the sensors, then clearly the question you are raising is a problem.

Gellie: Have you then looked at the mechanism for increasing the sampling rate in certain circumstances?

MacLeod: We have not specifically looked at a variable sampling rate.

Gellie: It would seem to be desirable to have a dynamically controllable sampling rate.

Kopetz: This is all getting back to a point which we implement in our system. It is clearly application dependent and we feel that we should give the user tools which allow him to configure the best system. This is why we introduce our concepts of immediate and consistent state messages. There are situations in which you would like to be consistent and there are other situations in which speed is important, and I think that we can only decide which is appropriate in the context of the application.

Byrne: It appears to me that we have to become very neurotic and suspicious! Every time a change occurs we must look at least once more before we really believe they have changed! If, however, you had a situation in which you have two observer processes which essentially have memory images you would be able to detect a change very rapidly. Would it not speed the system up considerably if you only looked a second time if there was disagreement in the arrays?

MacLeod: This is possible, but of course the arrays are not globally available!

Sloman: Returning to the question of tactical and control decisions, Professor Kopetz mentioned that he wanted a common image data base. However, there is a problem that in a lot of industries you are forced by law to have completely independent things like plant shut-down routines which are independent of the control. In such applications you clearly have to have separate data bases.

Kopetz: The question is, should the data base be created from the same sensor or different sensors? If it is from different sensors the it is a different system and you won't have this problem of integration. If it is in the same sensor base, then it is a different view of the system and although we would like to give this consistent image set we would agree basically that it is difficult to obtain it on a global context. In this case we would have to be happy if we can get it in the local context which is needed for decision making. In other words, the subsets of the main data base must be consistent since this is what is needed from the practical point-of-view. To get an overall consistent view appears to us to be somewhat difficult, but what is needed is consistency of the subsets which you are using to base decisions on and I think that this must be enforced.

Heher: I would like to raise another issue and that is the question of what happens under fault conditions. Typically, we don't just have a single signal change in state but a whole series of them which might carry on for an extended period. If then, in an image subset, we have such a train of variant signals coming down (which might easily be following each other two samples apart) and if we keep on delaying two samples before we transmit the data, then we can go for a long time with the data base in a consistent state. What would occur under these circumstances? Would we have to wait for the whole train of changes to be transmitted and received?

Kopetz: In this situation you are typically in the position where there is no consistent image set available. Everything is changing. In order to handle this I think you need to use what we call "immediate" messages. In this situation you are not really concerned about consistency - if a value goes beyond a certain limit and you are sure that this is far above what it should be, then you are not concerned about consistency, but you are concerned about speed. This is clearly an exceptional circumstance but one which is important.

Heher: I would not consider this, in fact, to be an exceptional case since it is one that is easily handled by relay-based shut-down systems because they are processing in parallel. It appears that you are saying we cannot use distributed PLC based shut-down systems?

Kopetz: I am not saying that you can't use them. I am saying that in such a situation then the requirement of consistency is important, but the requirement of speed is higher.

LaLive d'Epinay: If you define consistency correctly then you could still have a consistent state representation during a crisis period. For example in a big power distribution system you never find an interval in which there are no switches changing somewhere.

Brown: I think the problem with consistency must still be the analog values. This could be solved in the same way as you suggested, but then one must define the states and the quanta of analog values which make up a particular state. For a fast moving system you have to enlarge your quanta value and in a slow moving system you could decrease this value and you could have a trade-off with sampling time.

Rubin: One thing we know about an analog measurement is that it is always wrong! Therefore when we are talking about the consistency of an analog value, I think we are in a very complex domain and my approach would still be to use the Kalman-Gauss approach. In other words, we should rather talk about the best estimate in the case of an analog signal.

IEEE PROJECT 802: LOCAL AND METROPOLITAN AREA NETWORK STANDARDS (An April 1983 Status Report)

T. J. Harrison

IBM Corporation, Boca Raton, FL 33432, USA

Abstract This paper provides a status update on the IEEE 802 Project. It should be considered in conjunction with the author's previous DCCS papers on this subject (Ref.1,2) since it concentrates on changes since the last report. Changes in organization of the committee and the documents' method of processing as IEEE standards, and technical requirements are discussed. In addition, an expanded scope of committee work is described. The status of the work in the IEEE and international standards organizations is described.

Keywords. Project 802; local area network, metropolitan area network; LAN; telecommunication; token passing, CSMA/CD; OSI; reference model; communications; computer applications.

DISCLAIMER

It is prudent to remind the reader of several cautions and conventions that apply to the current status of Project 802 and, in fact, to the tentative results of any standards developing organization:

> This paper is not an official publication of IEEE Project 802 or of any of the other organizations discussed, and it has not been approved by these bodies. It is based on information publicly available in the literature, meeting minutes, and working papers of the committee. The most recent meeting was held 1983 March 20-25 and minutes and corrected working documents from this meeting were not available at the time of writing this paper. As a result, information relating to decisions made at that meeting are based on the author's notes.
>
> Opinions and speculation as to the content of the final standards, the current group consensus, and future actions are those of the author and do not necessarily represent the position of the committee, its members, or their sponsoring organization.
>
> The proposed standards have not been approved and all are subject to changes as they continue through the development and approval process. This paper and the tentative standards should be used with great caution with respect to decisions affecting future product design, specification, sales activity, or planned use of local or metropolitan networks.
>
> The author is not a member of the IEEE Committee but has been indirectly associated with its activities through involvement with a number of national and international standards organizations.

INTRODUCTION

This is the third paper on this subject that the author has prepared for the IFAC Distributed Computer Control Workshops (Ref.1,2). In proposing this paper for DCCS'83, the author suggested that perhaps it should be subtitled "The Last Status Report," and perhaps it will be. However, recent changes in the scope of the IEEE 802 Committee indicate the committee will continue work in this area.

The purpose of the paper is to provide an update on the activities of this very important and active committee. The emphasis is on changes made in the proposals since March 1982, changes in the committee structure and organization, and the current status of the work. For the reader unfamiliar with the proposed standards, it is strongly suggested that the previously cited references or other literature be used in conjunction with this paper. For ease of understanding, however, a brief resume of the contents of the standards is provided.

In addition to the IEEE 802 Committee, other organizations have been developing plans and proposals in areas directly or closely related to the 802 activity. Specifically, the PROWAY activity of IEC/SC65C/WG6 appears to be moving into closer alignment with 802 concepts and ECMA (European Computer Manufacturer's Association) is processing proposals that are essentially identical to those of the 802 Committee. These activities will be reviewed briefly in this paper.

Organizational Changes

Significant changes in the organization of the 802 Committee are affecting both the format of the standard documents and their processing. Since its inception in 1980 February through the summer of 1982, the committee had been composed of a plenary body and three subcommittees on Media, Data Link and Media Access Control (DLMAC), and High Level Interfaces (HILI). It was a functional organization closely aligned with the organization of the Open System Interconnection Reference Model (OSI/RM)(Ref.3). The organization of the proposed standard was similar, being a single document with sections corresponding to the subcommittees (although at that time the functions of the HILI subcommittee were not well defined and it did not have responsibility for a particular section of the standard).

With this organization, all proposals of the subcommittees required acceptance by the plenary body consisting of about 100 people. In addition, the concept of a single standard implied that none of the different systems (primarily differences in access method and media) could be approved until all were decided upon. Yet, in March 1982, the degree of completion of the various parts of the proposed standard was quite different. As a result, it appeared that there might be a significant delay in the final approval process and publication of the standard. The requirement for plenary approval meant that some highly technical details were being debated in a group of 100 people, some of whom were not interested or qualified in the debate.

In recognition of these problems, Maris Graube (802 Committee Chairman) proposed a committee reorganization and the separation of the single proposal into a "Family of Standards," with each member of the family being relatively independent of the others so as to allow independent approval. The proposal, not without controversy, was accepted by the committee. The current committee organization is shown in Fig.1. The Executive Committee consists of the chairmen of the six subcommittees (the Metropolitan Area Network subcommittee is relatively new since the last report and will be discussed later).

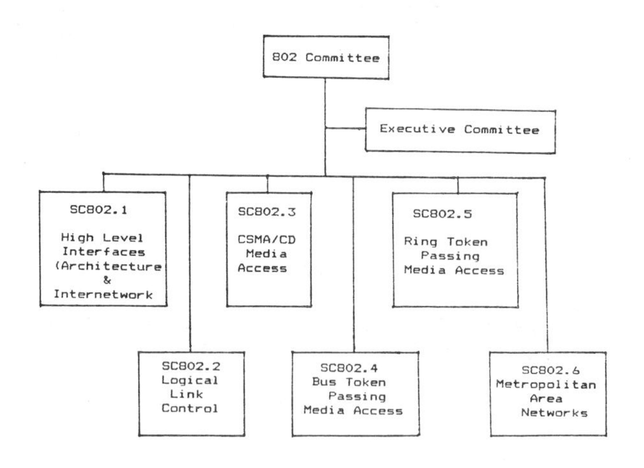

Fig.1: IEEE 802 Committee Organization

The Executive Committee is responsible for the overall guidance of the work and for approving submittal of proposals to the project sponsor for the next step in the approval process. Such proposals are examined for their readiness and further processing on the basis of the procedures that have been used by the subcommittees (e.g., resolution of subcommittee negative votes, outstanding issues, etc.). In general, the Executive Committee is not ruling on technical content as this is considered the responsibility of the subcommittee.

The scopes of the subcommittees are as follows:

<u>Subcommittee 802.1: High Level Interfaces:</u> This SC is responsible for determining the overall architecture of the standards document and for investigating the requirements for interfacing to higher layers in the OSI/RM. Specifically,

they are developing a document describing, as a guideline, the overall relationship between the other standards in the family. In addition, they plan to develop a standard dealing with internetworking, network management, and addressing.

Subcommittee 802.2: Logical Link Control. This SC is responsible for defining a standard for logical link control (LLC). The LLC is defined as a sublayer within the OSI/RM Layer 2 (Data Link Control). The proposed standard dS802.1 will be a prerequisite to the standards defining the media access control methods and also to the Metropolitan Area Network standard.

Subcommittee 802.3: CSMA/CD Media Access Method. This subcommittee is developing a standard which describes the Carrier Sense Multiple Access (with Collision Detection)media access method, known as CSMA/CD, for controlling use of the physical channel between multiple stations. This represents a sublayer of the Data Link layer of the OSI/RM. In addition, the subcommittee is defining the physical layer and the media requirements associated with the CSMA/CD access method.

Subcommittee 802.4: Token Passing Bus Access Method. The standard being developed by this subcommittee implements the token passing media access method on a bus topology. It also describes specifications for a corresponding physical layer and media.

Subcommittee 802.5: Token Ring Access Method. This subcommittee is responsible for a token passing media access method to be used on a physical ring, with the corresponding requirements for the physical layer and media.

Subcommittee 802.6: Metropolitan Networks. This recently formed subcommittee is responsible for defining a network suitable for use in a geographic region greater than that envisioned for the local area network, such as several city blocks or buildings.

In addition to these subcommittees, three technical advisory groups have been formed to study relevant technologies. Currently under study are fiber optic media, broadband transmission systems, and performance and modelling techniques. These studies may result in additional standards, or may merely support the work of the existing subcommittees.

Several other organization-related changes have also been made. Since its inception the committee has been officially sponsored by the IEEE Computer Society. Recently, the Computer Society assigned sponsorship to its Technical Committee on Computer Communications (TCCC), a group of about 190 people. The significance is that the operating procedures of the IEEE require a 75% affirmative vote of 75% of the members of the sponsor before a proposed standard can be forwarded to the IEEE Standards Board for final approval. Since the membership of the Computer Society numbers in the tens of thousands, meeting the 75% voting requirement and responding to a potentially large number of comments would be an almost impossible task for the committee. Generally speaking, the members of the TCCC are particularly interested in communication topics and are a representative forum of those interests substantially concerned with the topic of the 802 standards.

The result of these changes has been a change in the manner in which the committee meetings are conducted and in the way the 802 standards are being processed for approval. The meetings now consist of an opening plenary session of half- day for reports of the subcommittees and technical advisory groups and other matters of general interest. This is followed by 3 1/2 days of subcommittee deliberations. A final half-day plenary session is held to report the results of the subcommittee work. Subcommittee documents are no longer subject to an acceptance vote in these plenary sessions. Rather, each subcommittee conducts a 30-day ballot when it believes its work is ready for further approval. A majority vote of 75% of eligible committee members, coupled with resolution of negative votes, is required. When obtained, the proposed standard is forwarded to the 802 Executive Committee with a recommendation that it be forwarded to the TCCC for the next step in the approval process. As noted earlier, the Executive Committee is primarily concerned with the procedural aspects of the subcommittee work and the overall quality of the document.

The document then is circulated to all TCCC members for vote and comments. As noted above, a 75% plurality of 75% of the TCCC membership is required. Mandatory comments and comments accompanying negative ballots are returned to the 802 Committee (actually to the responsible subcommittee) for consideration and resolution. If the changes resulting from the comment resolution process are deemed not to be substantial, the committee, after editorial revision, may recommend to the TCCC that the document be forwarded to the IEEE Standards Board. On the other hand, if substantial changes are made, another vote of the TCCC membership is required.

Once forwarded to the IEEE Standards Board, the proposed standard is examined by the Review Committee (RevCom) of the Board to determine that IEEE procedures have been followed, that the necessary liaison has been effected, and that comments have been handled in a thorough and fair manner. On rare occasions, RevCom may raise technical issues but, in general, their review is procedural in nature. When satisfied, RevCom recommends acceptance or rejection to the full IEEE Standards Board. If the Board approves the recommendation of RevCom, the proposed standard becomes an IEEE Standard and usually, but not always, is forwarded to the American National Standards Institute (ANSI) for consideration as an American National Standard under the accredited organization approval method. The decision on proceeding toward international standardization theoretically is made by ANSI. In actual fact, it is based on the recommendation of the organization to whom ANSI has assigned responsibility for the applicable international committee.

Changes in the Document Format

A second significant change from the organizational change and decisions of the is that the work of the committee will result in a "Family of Standards," described as a set of separate standards which can be processed through the approval procedure as independent documents. Although it is intended that the standards be used in conjunction with at least one other member of the set, each subcommittee is able to proceed at its own pace in the approval process.

The set of standards in the family and their relationship is shown in Fig.2. The 802.1 proposed standard "Architecture and Internetworking" serves a dual purpose. The architectural portion of the document will not be mandatory, but rather a guideline which describes the interrelation between the other standards in the family. It also suggests how to use the standards. Since it is not a standard, it is shown dotted in the figure. The second part of 802.1 will be a standard concerned with inter networking, network management, and addressing. This will encompass the requirements necessary for a station in one local or metropolitan area network to effectively communicate with a station in another

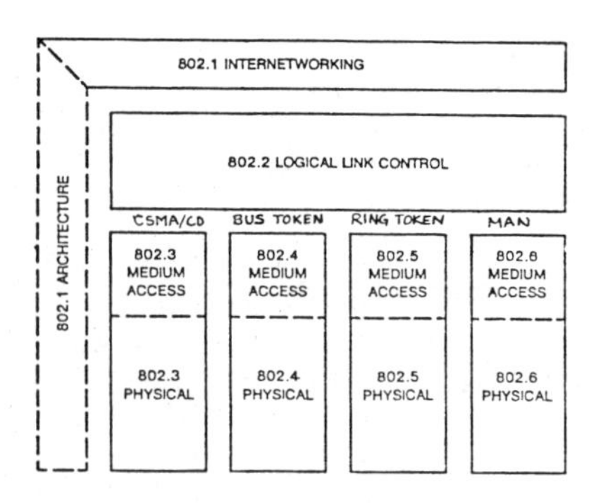

Fig.2: Relationship Among IEEE 802 Standards

network. Included will be the question of unique addressing. These questions are currently under intensive study and will not be included in the initial version standard.

The proposed standards 802.2 through 802.5 specify frame structure, protocol, and other requirements that correspond to Data Link layer requirements which are independent of the media access method. Although the document has been substantially revised in the past year, its scope and content have not changed substantially. It is intended that 802.2 will be a prerequisite to the 802.3, 802.4, 802.5, and 802.6 standards.

The proposed standards 802.2 through 802.5 specify requirements for the media-dependent functions of the Data Link Layer and the Physical Layer, and the media itself (which is not included in the OSI/RM). In each case, the proposed standard for the physical and media layers specifies one or more signal encoding and modulation techniques, media, and protocol for interaction with its corresponding MAC sublayer. In several cases, the standard provides detailed specifications for only one or two media while indicating that future standards will be developed for additional media. Thus, it is anticipated that the 802 family of standards will continue to grow.

SUBCOMMITTEE OVERVIEW AND STATUS

In the following sections, each of the subcommittees and their developments are briefly described. In addition, their status at the end of the March 1983 meeting and their future plans are described. For brevity, the abbreviation SC802.# is used to refer to the subcommittee and dS802.# is used to refer to their draft document.

SC802.1 Architecture and Internetworking

This subcommittee is a restructuring of the previous High Level Interfaces (HILI) subcommittee with a better defined purpose and responsibility for a separate document. Specifically, SC802.1 now has responsibility for the draft standard dS802.1, entitled "Architecture and Internetworking". This document has a threefold purpose: (1) To explain the relationship between the other standards in the "802 family" (This portion of dS802.1 is not mandatory, but rather is a guideline or tutorial; the introductions to the various sections in the previous version of the 802 Standard - Draft C.); (2) To provide a definition of "compliance"; that is, the conditions which must be met for equipment to be considered as complying with one or more of the proposed standards; And (3), to provide standards for internetworking, addressing, and network management. Although work on these latter three subjects has just begun and will not be included in the first version of the approved standard, their inclusion in the scope and Table of Contents of dS802.1 serves notice that they will be addressed in the near future.

Much of the material in the initial sections of dS802.1 was previously contained in one or more of the introductions to the sections in previous versions of the draft standard. It includes a definition of a Local Area Network, the objective of the standards, the relationship between the various standards, and the relationship to the ISO Open System Interconnect Reference Model (OSI/RM) (Ref.3). In addition, however, these sections now include a definition of a Metropolitan Area Network (MAN) and its relation to a LAN.

The definition of a Local Area Network has not changed substantially in the past year. The current definition, with added words underlined and deleted words in parentheses, is:

> A Local Area Network (LAN) is distinguished from other types of data networks in that the communication is usually confined to a moderate size geographic area such as a single office building, a warehouse, or a campus. The network can <u>generally</u> depend on a communication channel of moderate to high data rate (which has a consistenly low error rate), <u>low delay, and low error rate.</u> The network is <u>generally</u> owned and used by a single organization. This is in contrast to (long distance networks) <u>Wide Area Networks (WANs)</u> which interconnect facilities in different parts of the country or are used as a public utility. The local network is also different from networks which interconnect devices on a desktop or components within a single piece of equipment.

The addition of Metropolitan Area Networks to the scope of the 802 committee resulted in the addition of a definition for such networks:

> A Metropolitan Area Network (MAN) encompasses a larger geographic area such as several blocks of buildings. As with local networks, metropolitan area networks can also depend on communication channels of moderate to high rates. Error rates and delay may be slightly higher. A metropolitan area network may be owned by a single organization, but will usually be used by many individuals and organizations. MANs may be owned and operated as public utilities. They will often provide the means for internetworking of local networks.

Later in the document, the last sentence of this definition is strengthened to state " ... MANs often will provide the <u>primary</u> means for achieving inter-LAN communication" <u>Author's emphasis</u>.

With the addition of MANs, the scope of the standards is now stated to be "The scope ... encompasses a MAN standard and internetworking among LANs and MANs." This is, in the author's opinion, a significant change in that the committee seemingly was reluctant to address internetworking during its early history. There seemed to be a feeling that LANs would be used in isolation and considerations related to transferring data between networks were either unnecessary or beyond the scope of the committee.

The list of applications to be supported by the proposed standards remains the same (file transfer and access protocols, graphical applications, word processing, electronic mail, remote data base access, digital voice) except that "digital video" has been added. The example data devices to be supported remain exactly the same: computers, terminals, mass storage devices, printers/plotters, photo and telecopiers, image monitors, monitoring and control equipment, and gateways to other networks. The lists are stated to be non-inclusive.

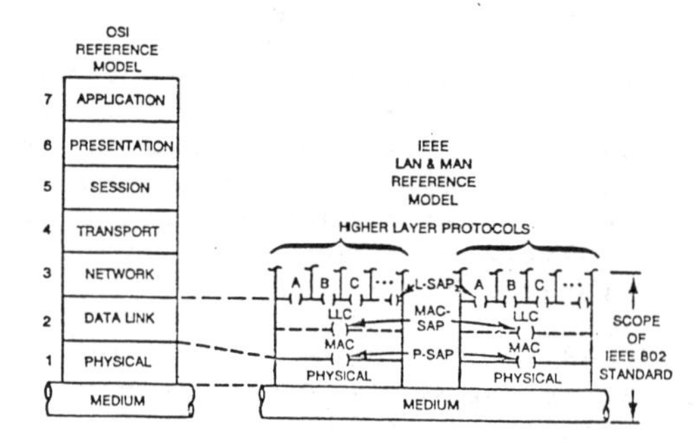

Fig.3: Relationship of Reference Models

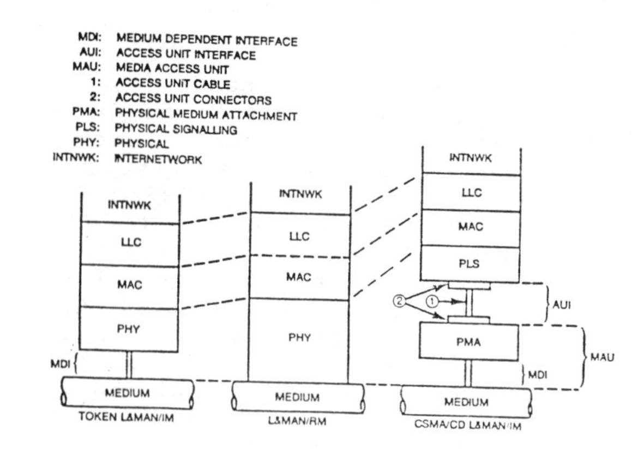

Fig.4: Implementation Reference Models

The proposed 801.1 standard includes a short tutorial on the OSI/RM and relates the reference models for LANs and MANs (L & MAN/RM) to this overall model. This relationship is illustrated in Fig.3. As before, 802 primarily is concerned with the physical and data link control layers of the OSI/RM. The LAN/RM and MAN/RM split the data link control layer into a Logical Link Control Layer and a Media Access Control layer. In addition, it explicitly includes a "media layer" which is only implicitly included in the formal OSI/RM. Due to the requirement for internetworking, 802 is also concerned with higher layers in the OSI/RM, but only to the limited degree that they affect internetworking.

The committee also has defined some intra-layer interfaces as shown in Fig.4. These interfaces, particularly in the case of CSMA/CD, define boundaries that may be accessible in some physical implementations. Due to the differences in the media access methods, these intralayer interfaces are unique to each access method.

The section on compliance defines the criteria under which a particular device or system can be claimed as conforming to the standard. The standards in the 802 family contain a significant number of options, some of which are only "recommended" or "preferred." This can pose a problem when selecting equipment which is to operate with other equipment from another manufacturer or of another vintage. The dS802.1 standard recognizes four cases and defines compliance requirements for each of them:

1. <u>A full-standard Media Access Unit (MAU)</u>

   a. Must provide all mandatory functions of a standard MAU
   b. Must operate at one or more of the standard data rates
   c. Must have a standard Access Unit Interface (AUI), and
   d. Must have one of the standard Media Interfaces (MI)

2. <u>Device with only the DTE side of the AUI</u>

   a. AUI must be exposed
   b. Must operate at one or more of the standard data rates,
   c. Must have the mandatory logical link control (LLC) capabilities, and
   d. Must use one of the standard media access methods

3. <u>Device where AUI is not exposed (i.e., integrated MAU and DTE)</u>

   Must meet all requirements of above item <u>except</u> those relating to the AUI.

4. <u>Half-standard MAU (i.e., custom medium, standard AUI)</u>

   a. MAU must provide all mandatory MAU functions, and
   b. Must operate at one or more standard data rates

Although this appears relatively stringent, many crucial parameters for interoperability (e.g., data rates and frequency band allocations) are, in fact, only recommended or stated as "preferred." In addition, the compliance requirements do not indicate how, if at all, a device is to be marked or labeled to document compliance.

The first four sections of dS802.1, essentially discussed above (Introduction, Reference and Implementation Models, Compliance, and Service Primitives), are considered by the SC to be ready for a SC vote, although the area of compliance is an admitted problem area. The remaining four sections (Network Management, Internetworking,

Addressing, and Further Study (transport issues)) are currently under study and will not be included in the initial version.

At the end of the March 1983 meeting, it was stated that there is consensus in the SC on the general architecture for network management, although it was not described in detail. For addressing, the preferred approach is to use a "flat" 48-bit address; i.e., the address will not be segmented in the standard to provide for a hierarchical scheme. On the other hand, it is recognized that the hierarchical approach is used and preferred in wide area networks (e.g., the country code, area code, exchange, number approach of the telephone network). Thus, a means to support the 48-bit flat address of the LAN in a WAN must be "invented."

In summary, it appears that those portions of dS802.1 that currently exist are relatively firm and are likely to be included in an initial approved standard on a schedule approximately the same as the other standards in the family (with the exception of dS802.6 on MANs). This will be helpful to early implementers. of the standard. One must be concerned, however, at the lack of definition in the areas of network management, internetworking, and addressing. The approval of dS802.2 through dS802.5 prior to these sections being available could result in problems, major revisions, or awkward solutions at some time in the future.

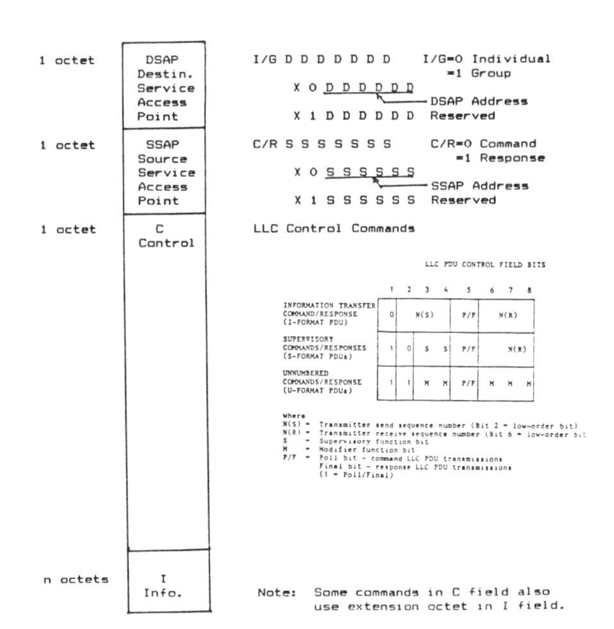

Fig.5. Summary of Logical Link Control (LLC) Frame

SC802.2 Logical Link Control

Logical Link Control (LLC) was not described in detail in the previous papers presented at the DCCS Workshops in 1981 and 1982, nor will it be here. It is an area that has not been controversial in the 802 committee. It is heavily based, at least conceptually, on previous protocol standardization activity; viz. the Asynchronous Balanced Mode (ABM) protocol of ISO 6256-1979 and ANSI X3.66-1971 (Ref.5,6), upon which the CCITT Recommendation X.25 Level 2 LAPB procedures are defined. Many of the commands and responses are, in fact, identical. The differences that do exist stem primarily from the fact that the 802 work assumes a peer-to-peer interchange whereas the previous standards assumed a so-called "master-slave" relationship.

The 802 LLC provides for two types of data link control operation as a means of satisfying the broad range of applications envisioned. Type 1 service, described as connectionless, is a minimum complexity protocol which may be useful when higher layers provide the necessary recovery and sequencing services or where it is not necessary to guarantee delivery of every Data Link Layer data unit.

Type 2 service provides Type I function and, in addition, provides a connection-oriented service comparable to those provided in existing national and international standards such as ADCCP and HDLC. This enhanced service provides guaranteed, sequenced delivery of Data Link Layer data units and error recovery techniques.

A third service type has been considered but is not included in the current draft. It is called "connectionless with immediate response" and is informally designated as Type 1.5. It is considered useful in some applications and the PROWAY group (discussed later) considers it to be essential in process control applications. The subcommittee has agreed to study it and it may be added to a future revision. It should be noted that the MAC function of the ring token access method, as discussed later, provides a form of response through the combination of the "Address Recognized" and "Frame Copied" bits in its frame status field.

The LLC data protocol data unit consists of four fields as shown in Fig.5. This has not changed since last year although there has been some "bit twiddling" in the assignment of codes in the Control (C) field.

SC802.2 Status and Plans: The activity of the past year primarily has been concerned with restructuring the document and providing a formal description of the LLC protocol. A Draft D document (Ref.7) was sent to the TCCC for ballot in early 1983. Although the ballot period was extended to 1983 May 15, the preliminary ballot result at the time of the March meeting was 86% favorable response from 45% of the TCCC. The March meeting was devoted partially to resolving the comments received to date. At the end of the meeting it was reported that no significant technical changes were made in responding to the comments. Most responses involved clarifications and rewording to minimize ambiguity. Nevertheless, the subcommittee expects to add appendices to the document and has decided that a revised document should be reballoted by the TCCC. It is hoped that this can take place in about July, with the goal of reaching the IEEE Standards Board for final approval at their 1983 December meeting.

SC802.3 CSMA/CD Access Control Method

This access method was reasonably well defined (Ref.8) early in the 802 work and is largely based on an existing design, so its description has remained relatively stable. The basic idea for controlling media access is that each station listens before transmitting to determine if the line is clear. If it is not (i.e., energy is detected on the line), the station defers until the line has been idle for a predefined length of time to ensure that other stations and the line have recovered. Once the station begins transmitting, it continues to listen for a period of time, called the "collision window", to ensure that a "collision" has not taken place. A collision occurs if a station begins transmitting when, due to propagation delay, it is unaware that another station has already started a transmission. The length of the collision window essentially is equal to the maximum roundtrip propagation delay plus station delays. A related parameter, equal to this delay plus the maximum "jamtime" (described below) establishes the maximum medium acquisition time and the maximum frame fragment size resulting from a collision. It is used as a parameter in scheduling retransmission attempts.

The need for the station to remain active for at least the duration of the collision window establishes a minimum frame size, giving rise to a requirement for the LC (Length Count) (See Fig.6) and optional Pad fields in the frame format, as discussed below. The minimum frame size is dependent on both the transmission rate and the physical extent of the network. In the recommended implementation at 10Mb/s, the minimum frame size is 512 bits or 64 octets. Allowing for the DSAP, SSAP, and C fields in the LLC data unit, this requires that the sum of the actual Information field and Pad field be a minimum of 46 octets or 38 octets, for 2-octet or 6-octet station addresses, respectively.

If a collision is detected, the station reinforces the collision for a brief period (the "jamtime"; 32 bit times in the 10Mb/s implementation). This ensures that all stations detect the collision. Each transmitting station then quiesces and delays another transmission attempt by a pseudo-random time. The delay is an integral multiple of the slot time. The integer is a uniformly distributed random integer r in the range 0 to 2**k, where k is the number of the retransmission attempt up to a maximum specified backoff limit. In the recommended implementation, a station makes up to 16 attempts (the Attempt Limit), using successive integers 1 to 10 (the Backoff Limit). This approach causes the delay to double each time up to 10 attempts, after which it remains constant at 10 slot times. If unsuccessful on the 16th attempt, an error is reported to a higher layer. This technique of scheduling retransmissions is called "truncated binary exponential backoff".

The frame format for the CSMA/CD access method has not changed since last year and is summarized in Fig.6 where the data field is the LLC data unit consisting of the one octet DSAP, SSAP and Control (C) fields and the multiple octet Information (I) field. The maximum frame size in the recommended implementation is 1518 octets. No explicit frame End Delimiter is

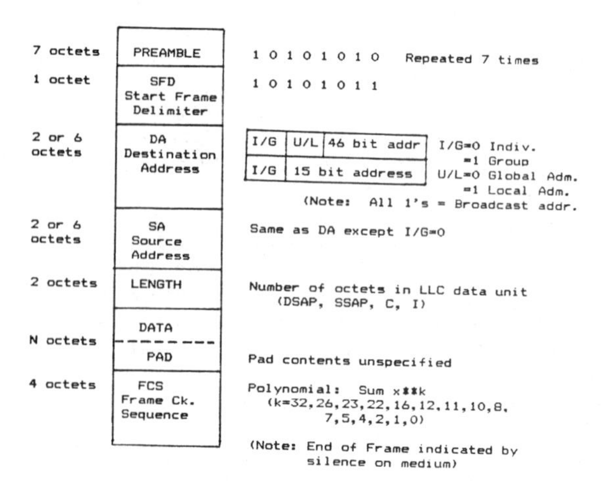

Fig.6: Summary of CSMA/CD Frame

required since the frame end is detected by lack of carrier energy on the medium.

Due to the nature of the technique, the collision detection circuits must be located very close to the transmission media, within 3cm in the recommended implementation. As a result, the standard an implementation consisting of a Media Access Unit (MAU) located at the trunk cable and connected to the Data Terminal Equipment (DTE) by four shielded twisted pairs for data and control (Data In/Out; Control In/Out; approximately 78 ohms characteristic impedance), a shielded twisted pair for power (voltage plus, voltage common), and a protective ground. The MAU may be powered through this cable which can be up to 50m in length. This configuration is implied in Fig.4 which explicitly shows the Access Unit Interface (AUI).

Two different signal encoding techniques are used at the AUI. Data (Data In/Out) is Manchester encoded, whereas control signals (control In/Out) utilize a frequency shift method. Signalling rates of 1, 5, 10, and 20Mb/s are provided for in the standard and the interface must support at least one of these rates (presumably the recommended 10Mb/s rate). Detailed specifications for the 10 Mb/s drivers and receivers are provided in the standard.

802.3 Physical Media. The current standard describes only a single medium and assumes operation at the 10Mb/s recommended rate. Other media and rates are being considered by the subcommittee for future revisions of the standard. The specified medium is 50 ohm coaxial trunk cable having an outside diameter of from 0.375 to 0.405 inches (0.95 to 1.0 cm), depending on the jacket and insulating material. A system is configured using 500m segments of cables which may be constructed of segments 23.4, 70.2, and 117m in length (odd multiples of half wavelengths at 5MHz) to minimize reflection at the cable connectors. A system may consist of up to five 500m segments coupled through repeaters, with each segment supporting up to 100 stations (MAUs and repeaters). The recommended cables are marked with an annular ring every 2.5m and stations are to be attached only at these points. This guarantees a minimum spacing and prevents alignment on fractional wavelength boundaries.

Although no specific connector is specified for the connection between the cable and the MAU, the standard seems to show a preference for a piercing tap connector (the so-called "vampire" tap) similar to that used in some CATV applications. The requirement is that the tap must provide less than 2pF nominal loading at 10MHz (4pF including attached active circuits).

System requirements are such that the cable can be grounded at only one point along the length of the cable. This leads to recommendations for non-removable "boots" or other insulating devices around connectors to prevent inadvertent grounding which can affect system performance. In addition, unless proper precautions are taken, single- point grounding can lead to safety problems due to the build up of static charge, high energy transients, or common mode voltages. The standard specifically addresses safety requirements and procedures to be used during installation and maintenance.

802.3 Status and Future Work. The current Revision D proposed standard (Ref.9) was sent to the IEEE Technical Committee on Computer Communications (TCCC) for a vote in early 1983. At the time of the last meeting (March 1983) 45% of the 192 TCCC members had responded, with 78% responding favorably, although many with comments. Although the balloting period was not to close until May 15, 1983, the working group considered and processed 67 comments at its March meeting. In their summary report, they reported that most of the comments caused "technical rewording" but no substantial technical changes were required. A meeting is planned for July 1983 to consider the additional comments received during the remaining ballot time. If no substantial changes result and if the TCCC final vote provides the necessary 75% plurality, the proposed standard will be forwarded to the IEEE Standards Board after some editorial work.[1]

The SC plans to include a comprehensive section on Network Management in a future revision. This will include traffic consideration, errors, maintenance, and planning with mandatory, recommended, and optional requirements. It will replace the brief and relatively general section in the current draft.

The SC is also considering two significant proposals for addition to the standard. The first, originating out of the work of the European Computer

[1] The 802.3 Standard was approved by the IEEE Standards Board on 23 June 1983.

Manufacturers Association (ECMA) is variously referred to as "Cheapernet" or "ThinNet" and a joint effort with ECMA has been established to consider the proposal.

The goal of Cheapernet is a less expensive CSMA/CD system that can be user-installed. It would provide for the use of small-diameter (approximately 0.2in (0.5cm) 50 ohm flexible coaxial cable (RG58/U) and standard BNC connectors. The connection to a combined MAU-DTE station would be made with a BNC "T" connector on the back of the station, eliminating the need for a remote, powered MAU at the cable. Cable length would be limited to 200m, with 2m station placement, and the number of stations would be limited to 70 stations per 200m cable. The effect on error rate has not yet been determined and some question remains on dc loop resistance which could cause a problem in detecting collisions. This system would have relaxed installation and grounding rules. At the March meeting, SC members were shown a system essentially providing these capabilities which is offered by the 3Com Corporation for the IBM Personal Computer (Ref.10).

The second major proposal is for a broadband CSMA/CD system. The SC heard a proposal (Ref.11) by representatives of Sytek, Intel, and General Instruments which incorporates the following major elements: (1) A transmission system which can be used up to distances of 30 to 50km at submegabit rates; (2) two physical layer implementations: phase-continuous FSK which is cost effective for low and medium speed broadband networks and duobinary AM/PSK similar to that used for the bus token access system; (3) HDLC framing with bit stuffing; and (4) a somewhat different algorithm to schedule retransmission after a collision. Although it might be possible to design a broadband system which matches the current baseband MAU interface, this proposal argues for an "optimized" approach which would not preserve this interface. However, the proposal has some elements (e.g., bit stuffing) which were previously rejected by the 802 Committee. The proposal was taken under study by the SC.

In the summary report at the end of the March meeting, the SC chairman indicated that the SC had decided to move toward an "optimized" broadband system that probably would not be compatible at the AUI.

Internationally, ECMA began considering the CSMA/CD proposals some time ago. They essentially have taken the CSMA/CD document and split it into three documents. A number of differences arose between the 802 and ECMA versions, partially because of time delays in the liaison between the groups, but also because ECMA was concerned about, and concentrated on, the safety issue. It was reported that these differences have been largely resolved and that the ECMA documents may be ready for a vote by the General Assembly (the "official approval") by June 1983. In addition, ECMA desires additional sections on repeaters, network management, and an installation guide. They also are studying a paper on an optical fiber version of the CSMA/CD system.

SC802.4 Bus Token Passing Access Method.

The bus token passing access method assumes a topology of trunks and branches in a relatively unrestricted configuration except that there may be no closed loops. The access concept is that a "token" controls the right of access to the medium; i.e., possession of the token gives the station the right to use the medium for some predetermined period of time. Since there is only one token under normal circumstances, only one station will be actively transmitting, although all will be listening. The token is passed sequentially among the active stations, thus forming a "logical ring". Ring maintenance functions built into each station provide for ring initialization, lost token recovery, new station addition into the logical ring, and other error recovery procedures. A form of priority is also a feature of the 802.4 bus token access proposal.

The last major changes and additions to the bus token access method took place at the March 1982 meeting and were described in the previous paper (Ref.2). Since then, considerable work has been done on the document to restructure it according to the new organization and to properly describe the operation and implementation of this method. However, there have been no major changes in concept (Ref.12).

The essence of the bus token access method is as follows:

Multiple stations are connected to a shared broadcast medium which generally has the topology of a highly branched tree. Through an initialization procedure, one active station is given the "token" and it may then use the medium to transmit to any other station connected to the bus. The length of time that the token may be held, however, is constrained by a token-holding timer and the station must complete its transmissions before the expiration of the timer. When the transmission is completed or the token- holding timer has expired, the token is passed to another station for its use. In contrast to CSMA/CD, multiple frames may be sent during a single transmission.

The order in which the token is passed from station to station is established in the initialization procedure. This creates a logical ring in descending order of station address. The sequence of token passing is maintained by each station knowing its predecessor address (_i.e.,_) the address of the station from whom it receives the token) and its successor address (_i.e.,_ the address of the station to whom it passes the token).

It is important to note that the token holding station can communicate with a station which is not a part of the logical ring. This will be explicitly discussed in an appendix of the next revision of the proposed standard. Such a secondary station might, for example, have minimal function or not be equipped to handle the token. The communication with such secondary stations must be completed within the token-holding time and must not confuse or otherwise cause a malfunction of any other station.

The 802.4 bus token access method incorporates a priority scheme. It is called a "timed-token bandwidth allocation" method and, in essence, guarantees a certain percentage of the available bandwidth to stations requiring the highest class of service. The remaining bandwidth, and that not used by the highest class stations on a particular rotation of the token around the logical ring, is shared among stations in lower classes. Any number of classes is theoretically possible but the standard provides four such access classes. Implementation of the priority system is optional.

This type of priority system differs from a true preemptive system, such as that used in the ring token access method described later, in that stations of a lower class are not necessarily "locked out" even when high priority traffic is pending at a higher class. Some portion of the bandwidth is always available for sharing among the lower class stations.

The bus token frame structure has not changed significantly since March 1982. It is shown in Fig.7, where the MAC data unit is the LLC protocol unit consisting of the DSAP, SSAP, and Control (C) octets plus the Information (I) field consisting of multiple octets. The number of octets between the Starter Delimiter (SD) and End Delimiter (ED), exclusive, is a maximum of 8191. The Source Address (SA) and Destination Address (DA) may be either 2 or 6 octets but is the same throughout any given network at any given time. Conforming equipment is not required to provide both address lengths.

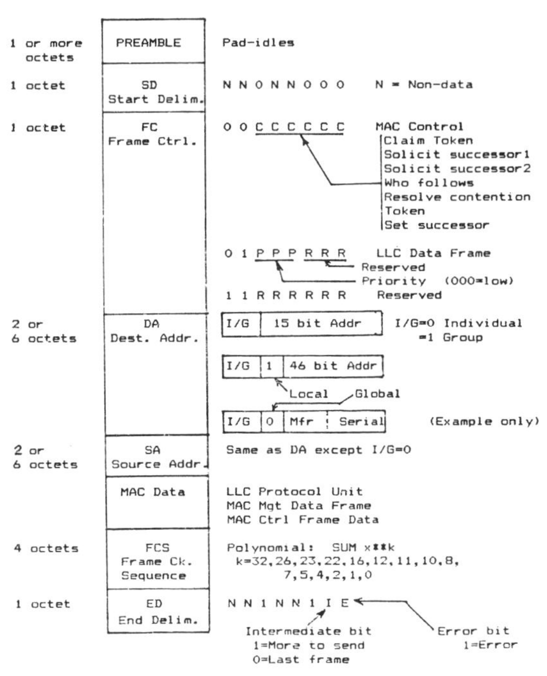

Fig.7: Summary of Bus Token-Passing Frame

802.4 Physical Media. Significant progress has been made in the precise definition of the physical media and signaling techniques since last year, although there have been no basic changes. The standard provides for three media: Two for baseband transmission and one for broadband.

The first provides for use of 75-ohm coaxial cable (such as RG6, RG11 or semi-rigid) at 1Mb/s and utilizes Phase Continuous FSK modulation. The recommended cable topology is a long unbranched trunk with short "stub" drop cables. The drop cable is a 35-to 50 ohm coaxial cable stub, less than 14 inches (35cm) long, and coupled into the main cable using a BNC "T" connector. Data is encoded using the Differential Manchester coding method with a high level being represented by a 6.25MHz signal and the low level being a 3.75MHz signal.

A CATV-like system, utilizing 75-ohm coaxial cable, is used with Phase Coherent FSK at data rates of 5 and 10 Mb/s. The recommended cable configuration is CATV-like semi-rigid cable with 75-ohm flexible drop cables. The drop to cable connection is made using a non-directional impedance-matching passive tap. Active repeaters are used for high-fanout branching and for extending the system beyond the basic signal loss limit imposed by cable characteristics. Data is encoded directly as an integral number of cycles of constant frequency. The frequencies used are 5 and 10MHz at the 5Mb/s data rate and 10 and 20MHz at the 10Mb/s data rate. A "zero" is represented by two full cycles of the higher frequency, a "one" is one full cycle of the lower frequency, and pairs of non-data symbols are represented by a three cycle sequence consisting of one cycle at the high frequency, one cycle at the low frequency, followed by one cycle at the high frequency.

The broadband system utilizes Multi-Level Duobinary AM/PSK encoding and modulation. The recommended cable configuration is a single 75-ohm CATV-like semi-rigid trunk with flexible drop cables connected by means of directional, passive, impedance matching taps. Since bidirectional transmission is utilized, the system uses a head-end regenerative repeater which receives on one frequency and retransmits on another. Although the single-cable system is recommended, the standard allows for the use of dual unidirectional cables. Standard CATV amplifiers can be used to provide extension of the system beyond the basic cable loss constraints. Required channel bandwidths are 1.5, 6, and 12 MHz at the 1, 5, and 10Mb/s data rates, respectively. The channel allocations are identical to those used in CATV activity (known as single-cable mid-split) with the recommendation that 89.75-92.25 and 282-283.5 be used at 1Mb/s, 83.75-89.75 and 276-282MHz at 5Mb/s, and 59.75-71.75MHz (two adjacent channels) and 252-264Mhz at 10Mb/s. These recommendations follow those of the EIA TR40.1 standards committee.

The duobinary encoding method utilizes two bits that specify the amplitude of the eventual modulation. The standard defines a single form, the 1 bit/Hz three-symbol signalling method, but includes compatibility considerations for three additional forms. In this method, a "zero" is represented as "0" which results in a zero amplitude modulation, a "one" is represented by a "4" which results in maximum amplitude modulation, and non-data symbols are encoded as a "2" corresponding to 1/2 maximum amplitude modulation.

802.4 Status and Future Work. The Draft D (December 1982) version of the token bus proposal standard was distributed to the TCCC for ballot prior to the March meeting and, at the time of the meeting, a 82% favorable response has been received from 47% of the TCCC members. During the processing of the comments at the March meeting, some changes were made in the document which are considered "substantial", although they do not affect any of the basic concepts. As a result, the SC intends to revise the document and submit it for another TCCC vote, hopefully in September 1983.

During the March meeting, the SC considered a request from the Instrument Society of America (ISA) standards committee SP72 which acts as the U.S. Technical Advisory Group (TAG) to IEC/TC65C/WG6, better known as the "PROWAY" group. It was reported (Ref.13) that SP72 began an evaluation of the 802 token bus and in January 1983 voted to recommend to WG6 that, subject to certain required enhancements to dS802.4, PROWAY be defined on a subset of the proposed standard. Earlier, in November 1982, the WG6 had expressed interest in such an approach. The SP72 recommendation requested several extensions to the current 802.4 system, including the "connectionless with immediate response" class of service. In total, there were eleven points which SP72 felt needed to be addressed. As a result of the presentation, SP72 will contribute an appendix to the 802.4 standard. This is planned for Draft E, due to be published in July 1983. It will be quite detailed and will include modified frame formats and finite state machine transition diagrams for the major enhancements.

802.5 Token Passing Ring Access Method.

The last major conceptual changes to the ring token access method took place at the March 1982 meeting where the priority scheme and several other major proposals were accepted. The basic requirements of the proposed standard have remained the same since then with the activity devoted to determining and describing the detailed requirements of the protocol and transmission medium.

Conceptually, the operation of the token ring is almost identical to that of the bus token, with media access determined by a token which is passed from station to station. The major difference is that the topology of the physical network is a ring, into which stations are inserted. Thus, it is a sequential medium, as compared to the broadcast medium of the bus token and CSMA/CD methods. The use of physical ring tends to simplify the token handling protocol in that the individual stations do not have to maintain explicit records of the predecessor and successor stations. In addition, the ring provides advantages with respect to error detection and recovery. It can also ease physical configuration problems since the signal is regenerated by each station, obviating the need for repeaters and amplifiers in many situations.

The operation of the ring is well described in the overview section of the proposed standard (Ref. 14). The following description is a slightly paraphrased version of that overview:

Access to the physical medium (the ring) is controlled by passing a token, indicated by a bit in the appropriate position in the Access Control (AC) field, around the ring. A free (i.e., non-busy) token gives the downstream (relative to the station passing the token) station the opportunity to transmit a frame or sequence of frames limited by a token-holding timer.

Upon request from an upper layer for frame transmission, the MAC function constructs a protocol data unit (PDU) from the data to be sent or from network management requests and prefixes the FC, DA, and SA fields (See Fig.8). The PDU is then queued awaiting the receipt of a non-busy token that can be used to transmit it. If the priority option is implemented, such a token must have a priority less than or equal to the priority assigned to the queued PDU(s). If a frame or token that cannot be used is detected, the station adjusts the AC field before repeating it to request a token of the appropriate priority.

When a usable token arrives, the station changes the token bit to busy, stops repeating the incoming stream, and begins transmitting the queued frame(s). During frame transmission the FCS is calculated and appended.

While transmitting the frame(s), the station checks to see if its own address has been returned in the SA field; i.e.,that its frame has traversed the ring. If it has not been seen by the time the end of frame sequence is transmitted, the station transmits fill characters until it has been returned. If, or when, its SFD is detected, the station may transmit a free token consisting of an abbreviated frame (SD, AC, ED) with the token bit in the AC field set to "0". Unless the ring latency is greater than about 150 bits, the token holder will not be required to send idles.

At this point, however, some of the token holder's transmission remains on the ring and the token holder must "strip" these bits. At the March 1983 meeting, it was decided that a single stripping method is to be employed.

Specifically, the use of the I (Intermediate bit in the ED) bit is mandatory and the token holder continues stripping until I = 0 (i.e., last frame) is detected. The station then returns to the repeating state.

At the receiving station, frames are detected by the receipt of a Starting Delimiter (SD). At the same time, the incoming bits are repeated on the outgoing line, delayed by a few bit times. The MAC-layer in the station decodes the destination address to determine if the frame is addressed to the station. If so, the MAC sublayer removes the AC, FC, and FCS, and passes the DA, SA, I, and class of service indicator to the Logical Link Control (LLC) layer.

The Address Recognized and Frame Copied bits in the Frame Status (FS) field are set to "1" before sending the FS to the next station. The combination of the A and C bits allow the originating station to differentiate among three conditions: (1) The station is non-existent or not active; (2) The station exists but did not copy the frame (e.g., its buffer is full); and (3) The station exists and copied the frame. In essence, this provides a form of the "connectionless with immediate acknowledge" class of service being considered by SC802.2 as Type 1.5.

If the frame is not destined for the station, the MAC-layer nevertheless checks the FCS and, if it detects an error, sets the error bit in the last octet before repeating it. Upon its return to the originating station, the token holder can only determine that an error took place. Presumably, however, a higher layer protocol (not considered in the standard) would either cause retransmission of the frame or make an inquiry to the receiving station. Alternatively, a higher layer in the receiving station could notify the sender of the error condition if it existed at the time of receipt. When such a message is not received by the originating station, it could safely assume the frame was received correctly.

The priority bits and the reservation bits in the AC (Access Control) field work together to match the ring service priority to the highest priority frame queued for transmission.

The current ring priority is indicated by the setting of the priority bits. When a station is not able to seize a token and has PDUs queued, it requests a token of appropriate priority by setting the reservations bits in the AC field. The value of the reservation bits may be increased by any station to match the priority of a queued PDU. This may be continued until a usable token is seized.

If a token holder receives a frame with the reservation bits set to a value higher than the current ring priority, the token holder, when it releases the token, sets the priority to the higher reserved value. When doing this, the station remembers the old service priority (say, Sx) and the new priority. The station(s) requiring the higher priority token now seizes the new token and uses it. Eventually, perhaps after several priority changes by other stations, a free token with priority Sx returns to the original token holder. Upon recognizing this, the station returns the service priority of the ring either Sx or the current requested value so as to eventually continue the original sequence of token passing.

This provides for a true pre-emptive system and assumes that some agreement exists by means of network administration or higher layers in the station as to relative priorities of PDUs from different stations. If this were not the case, every station could classify all PDUs to be the highest priority, negating the priority system. Of course, this requirement is common to all such systems; the existence of a "higher authority" is always assumed.

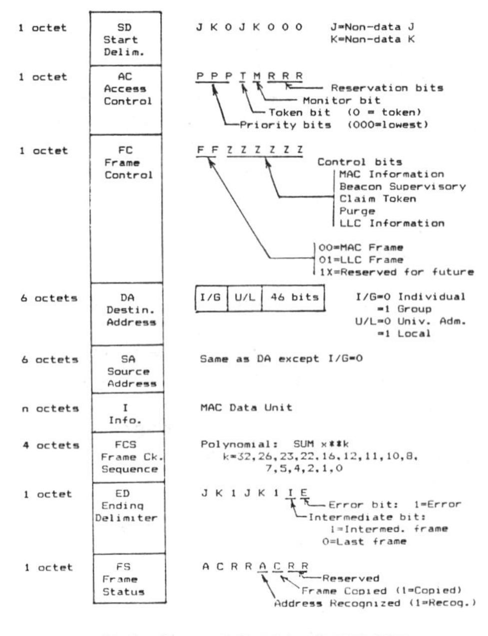

Fig.8: Summary of Ring Token-Passing Frame

The frame format is shown in Fig.8. It should be noted that, unlike the bus token and CSMA/CD methods, only a 6 octet address is allowed. This is a significant change that has taken place since last year. In addition, the use of the I (Intermediate Frame) bit in the End Delimiter (ED) field now is required as discussed earlier.

Procedures are provided for detecting and recovering from various errors. The ring method utilizes a centralized monitor concept whereby a single station is responsible for most of these procedures. This is a change from the previous drafts which utilized a distributed monitor function. Multiple stations may possess the monitor capability but only one, assigned by the initialization procedure, is active at any one time. It contrasts with the bus method in which the monitor function is distributed among all stations at all times. Should the primary monitor fail or be taken off line, defined procedures assign the monitor function to another station.

__802.5 Physical Medium.__ The proposed standard provides for a twisted shield pair as the trunk cable. Signal rates are 1 or 4 Mb/s using Differential Manchester encoding. Since the technique for connecting the station to the medium requires that the station be inserted "in series" with the line, the specifications require that each station ensure continuity of the loop while connected, but not active. This may be done by means of a bypass relay or equivalent mechanism and an example detection circuit is included in the proposed standard. If the station is actually removed, the specified connector automatically provides continuity of the pair by means of shorting contacts. The specified plug is hermaphroditic so that two identical units will mate when oriented 180 degrees with respect to each other. Thus, both the station and the cable port use the same connector. The shorting contacts which guarantee the loop continuity also provide an automatic looping capability for the disconnected station.

A second medium has also been discussed for the ring token system but is not included in the current draft. It would provide a coaxial cable baseband system at data rates of 4, 10 and 20 Mb/s. Due to the pressure to prepare and approve the initial standard, this proposal has not received significant attention during the past year.

__802.5 Status and Future Plans.__ Due to its later development schedule, the proposed 802.5 standard has not yet been forwarded to the TCCC for a vote. At the last meeting it was announced that all remaining technical issues at the SC level, including those raised by the corresponding ECMA group, had been resolved to the satisfaction of the subcommittee. In their review, ECMA requested a network design or installation guideline and the SC indicated that this would be forthcoming. The SC is also looking into the possibility of a formal protocol validation.

SC802.6 Metropolitan Area Networks.

This subcommittee is relatively new, having been formed during the past year as a result of the 802 scope expansion. As such, it has not yet produced a draft standard and many of the ideas only are at the conceptual stage. Essentially, however, the SC is seeking a system which would operate over a 50km diameter area at rates of 1 to 20 Mb/s.

Some of the desired characteristics include: (1) Provision for "grades of service", possibly by means of a priority scheme; (2) If a broadband system is decided upon, the channel allocation scheme in the country should be used; (3) Some degree of commonality with other networks is desirable. Lacking this, convenient internetworking should be possible; and (4) Media independence which would allow easy conversion to alternate media such as optical fiber and microwave is desirable.

Initially, a TDM scheme utilizing a CATV-like network was investigated. However, the discussion has now been broadened to include "conventional" schemes, such as the CATV-type, to satellite and packet radio transmission methods. Essentially, the SC is back at "ground zero" and is examining fundamental questions such as whether the network is public or private and the degree to which stations can be transported beyond the bounds of MAN.

With this expansion of the discussion, it appears that the SC may move in a direction that could violate some of the fundamental prior decisions of the 802 Committee, such as the idea of using a common LLC and some of the addressing decisions. Should this occur, the SC would need to seek permission of the Executive Committee and, potentially, the IEEE Standards Board. It is certainly too early to predict if this will happen but some concern has been expressed by members of the 802 committee.

CONCLUSION AND OBSERVATIONS

To many, the work of the 802 Committee has passed the exciting and fun stage and has settled into the detailed hard work that is a part of every standard development activity. It is the tedious work of "crossing every 't' and dotting every 'i'" that is so essential to the success of a standard.

The exciting initial brainstorming days are mostly gone. Also gone is the intense competition among access methods that characterized the early work. This latter absence, probably good at this stage, has been the result of three actions: (1) The reorganization of the committee no longer requires document approval of the entire committee, thus reducing the opportunity for advocates of the different media access methods to debate; (2) The decision to include all three access methods in the standard, thus creating a family of standards. This eliminated the "one or nothing" pressure evident in the early days; and (3) The accepted recognition that each access method has its advantages and disadvantages, its proponents and opponents, and that each may be preferable in a particular application.

This latter point is perhaps best demonstrated by the recently published results of a task force, commissioned by the 802 committee, which examined the maximum mean data rate for each access method under a variety of loading. This subject had been one of the main arguments during the early committee activity. The task force produced a 250-page draft report from which a summary article was produced (Ref.15). The summary paragraph of the article compares the methods as follows:

> The best available evidence, based on this study and related studies, is as follows. Token passing via a ring is the least sensitive to workload, offers short delay under light load, and offers controlled delay under heavy load. Token passing via a bus has the greatest delay under light load, cannot carry as much traffic as a ring under heavy load, and is quite sensitive to the bus length (through the propagation time for energy to traverse the bus). Carrier sense collision detection offers the shortest delay under light

load,[2] is quite sensitive under heavy load to the workload, and is sensitive to the bus length (the shorter the bus the better it performs) and to message length (the longer the packet the better it does).

Despite the competition and eventual resolution of questions such as this, the committee has thus far performed heroically and is on the verge of obtaining approval of a major standard in the relatively short time of 3+ years. Most observers seem to feel that the technical work is sound and some developers are proceeding to design products based on one or more of the proposed standards.

The committee has also changed its character and some of its initial goals in the past three years. Much of the pioneering work was done by a few individuals, only some of whom remain on the committee today. Many of the original members of the committee with a great desire for the standard, but little experience in the network protocol area have been replaced by skilled engineers with broad network and/or standards experience.

The goal of producing a single standard essentially has been abandoned; practically it remains in name only. The five sections (excluding 802.6 whose nature cannot be predicted) constitute, in fact, three different standards which, coincidentally, share a common logical link control definition (802.2) and a common introductory document (802.1). Products conforming to these standards will not be compatible in any sense and cannot coexist on the same network (it might be possible that they could coexist on the same media but no two standards currently specify the same media). Significant differences (e.g., 2-or 6-octet addresses in CSMA/CD and bus token but only 6-octet addresses in the ring token) have emerged. This is not to say that this is bad, only that it differs from the original goals of the committee in December 1980.

[2] Author's Note: Taken out of context, this statement is somewhat misleading in that shortest delay does not imply shortest in comparison to the ring and bus token-passing methods but rather in comparison to other loading conditions of the CSMA/CD method; i.e., delay in the CSMA/CD method monotonically increases with load. Under the assumptions used, however, the delay in the token ring is always less than in the CSMA/CD or the bus token access methods.

In the past several years, more than 100 LAN products utilizing some 35 different protocols have been announced by vendors. Some are related to the 802 work and attempt to anticipate the final form of the standards. Others have no relation to the work. And, more are being announced every week.

The 802 hoped to be an anticipatory standard, establishing itself before many significant products have been announced. In some degree it succeeded, since none of the 100 products seem to be the dominant choice of users. On the other hand, this degree of market activity is stiff competition for the acceptance of the 802 work. It is too early, in this author's opinion, to safely predict the eventual success of any or all the 802 standards. This is a question which will eventually be determined in the marketplace, as it should be. However, this author's feeling at the moment is that all three will become significant to some degree. Certainly there is a recognition that proper standards in this area are key to the widespread use of computers, and particularly microcomputers, in enterprises of the future. It is absolutely necessary that every device in such an enterprise have the capability of communication with every other device. These standards cannot guarantee that, since they do not deal with all the OSI/RM layers, but they will go a long way toward this goal. This recognition on the part of users and vendors is likely to be a key element in the eventual success of the standards.

REFERENCES

1. Gerald J. Clancy, Jr. and Thomas J. Harrison, "Local Area Network Standards -- A Status Report --" in W. Miller, Distributed Computer Control Systems, (Proceedings, Third DCCS Workshop, Beijing, PRC, August 15-17,1981), Pergamon Press, Oxford, England.

2. Thomas J. Harrison, "IEEE Project 802: Local Area Network Standard -- A March 1982 Status Report", in R. Tavast and R. Gellie, Distributed Computer Control Systems, Tallinn, Estonia, USSR, May, 1982. (Proceedings to be published by Pergamon Press, edited by R. Tavast).

3. "ISO/DIS 7498, Data Processing - Open Systems Interconnection Basic Reference Model", February 4, 1982.

4. "Draft IEEE Standard 802.1 - Architecture and Internetworking - Draft A", IEEE Subcommittee 802.1, March 1983.

5. "ISO 6256-1979," International Organization for Standardization. (Available through American National Standards Institute, New York).

6. "American National Standard for Advanced Data Communication Control Procedures," ANSI X3.66-1979, American National Standards Institute, New York, 1979.

7. "Draft Standard P802.2 Logical Link Control - Draft D", IEEE Subcommittee 802.2, November 1982.

8. "The Ethernet, A Local Area Network: Data Link Layer and Physical Layer Specification (Version 1.0)", jointly published by Digital Equipment Corporation, Intel Corporation and Xerox Corporation, September 30, 1980.

9. "Draft IEEE Standard 802.3: CSMA/CD Access Method and Physical Layer Specifications - Revision D", IEEE Subcommittee 802.3, December 1982.

10. "Etherlink", Form DC3C500/982/10M, 3Com Corporation, Mountain View, CA, undated.

11. "Proposal to IEEE 802 for CSMA/CD Broadband", Sytek, Intel, and General Instrument Corporations, (Distributed at March 1983 meeting of Committee 802), March 21, 1983.

12. Draft IEEE Standard 802.4: Token-Passing Bus Access Method and Physical Layer Specifications - Draft D", IEEE Subcommittee 802.4, December 1982.

13. "TC-5A Meeting Minutes, January 18-20, 1983, Philadelphia", International Purdue Workshop, Purdue Laboratory for Applied Industrial Control, W. Lafayette, IN 47907 USA.

14. "Draft IEEE Standard 802.5: Token Ring Access Method and Physical Layer Specifications - Working Draft:, IEEE Subcommittee 802.5, February 21,1983.

15. Bart W. Stuck, "Calculating the Maximum Mean Data Rate in Local Area Networks, Computer Magazine, May 1983, pp.72-76.

DISCUSSION

Kopetz: I would like to know whether PROWAY conforms to the 802 standards?

Harrison: My understanding is that it was designed with this standard in mind, although it may not currently conform exactly because there are still some changes occurring. In talking to people involved, my understanding is that you can, with some re-programming, conform to the standard.

Sloman: I wonder if Dr Harrison could explain why there are so many physical layers in the token-bus system?

Harrison: I suspect it was a practical consideration by the people involved, particularly as many proprietary interests were involved. In addition, I feel that there is a cost penalty to be paid for high speed. So if you look at the physical media you find that they are not all used at the same speed. The first one is a one megabit and the others five to ten and baseband while the other is broadband. So there is clearly no feeling today in the industry as to whether the ultimate answer lies in broadband or baseband transmission. I think it is clearly an effort to try and provide enough options and to allow cost trade-offs in some cases.

A FLEXIBLE COMMUNICATION SYSTEM FOR DISTRIBUTED COMPUTER CONTROL

M. Sloman, J. Kramer, J. Magee and K. Twidle

Department of Computing, Imperial College of Science and Technology, London SW7 2BZ, UK

Abstract. This paper describes an extensible communication system for the CONIC Architecture for Distributed Real-time Systems. The CONIC programming language primitives for both local and remote interprocess communication are presented. The communication system which supports these primitives is itself implemented in the CONIC language. It exploits the configuration flexibility of CONIC to provide a very simple basic datagram service which is extensible at configuration time to give additional services such as virtual circuit and multidestination. The paper relates the CONIC communication system to the ISO Reference Model.

Keywords: Interprocess communication, communication protocols, local area networks, network architecture, distributed control.

1. INTRODUCTION

Distributed Computer Control Systems (DCCS) consist of microcomputer stations, interconnected by a network. Using a communications system the stations exchange information to synchronise their activities and achieve their common goal of controlling machinery or an industrial plant. It must be possible to configure a DCCS to meet the needs of particular applications. In general, control systems do not remain static but evolve, and so the DCCS must be easily modifiable to incorporate new functions, modify old ones and generally tailor the system as required.

The CONIC Architecture [Kramer 83] provides a flexible environment for building and configuring systems from message passing modules. This paper shows how this configuration flexibility has been exploited in the design of the CONIC communication system, based on minimal functionality at the lowest level. The CONIC interprocess communication (IPC) primitives are powerful enough to support many process control applications and yet are sufficiently simple and efficient to be used to build special purpose modules. This has been demonstrated in the use of the CONIC IPC facilities for the implementation of the CONIC distributed operating system and communication system.

In order to set the requirements which the communications system must fulfill, we first outline the CONIC system and describe the IPC primitives in detail. The basic communication system is then described, giving the implementation for both local intrastation communication (provided by the local station kernel) and for remote interstation communication. The latter provides a basic datagram service with network routing. Services such as multidestination message passing, virtual circuits and information transformation are then described as simple configurable extensions to the basic system. It is also interesting in that both the basic interstation communication modules and the extensions are implemented using the CONIC programming language and IPC primitives themselves!

2. THE CONIC ARCHITECTURE

CONIC was developed to satisfy DCCS requirements [Prince 80]. The hardware structure consists of microcomputer stations connected to a network of multiple interconnected subnets (fig. 2.1), where the subnets are local area networks such as rings or serial highways. This allows extensibility of stations within a subnet and subnets within an overall network.

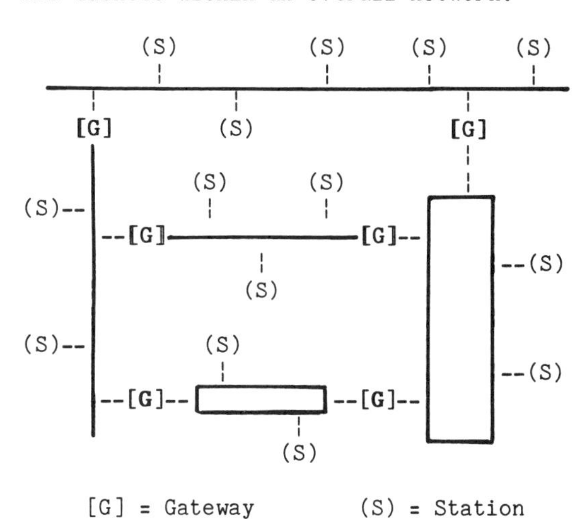

Fig 2.1 Interconnected Subnets

The run-time support in a station consists of a minimal kernel, written in Pascal, which provides multitasking and intertask (process) communication within a station. All other software in a station runs as application programs and is written in the CONIC Programming Language. This includes an executive to manage station resources and support dynamic configuration of the software and a layered communication system to support interstation communication.

The CONIC software architecture provides message passing IPC primitives and modules consisting of a set of concurrent tasks (processes). The approach clearly distinguishes between the programming of individual software components (module definitions) from system building and (re)configuration from instances of these modules [Kramer 83]. The programming of a module type is concerned with defining module behaviour and message **transactions**, whereas the configuration of a system is concerned with creating module instances at stations and setting up **connections** (associations) between these modules. We discuss system configuration first.

2.1 System Configuration

A **module** instance is the smallest replaceable software component in the configuration of a system ie. it is the software unit of distribution and must reside in a single station. It is possible, however, to have more than one module in a station.

Modules communicate by exchanging messages [Kramer 81]. The module interface is defined in terms of typed **exitports** and **entryports** which specify the message types which can be sent and received respectively. At compilation time, it is not necessary to know the other components with which a module will communicate as, inside a module, messages are sent to and received from local entities - the module exit and entry ports respectively. This makes it possible to program a module as a reuseable entity, where the same module may be used in many different configuration environments.

The configuration of a DCCS involves the creation of instances of modules at stations, and the **linkimg** of module exitports to entryports of other modules, eg. (figure 2.2)

```
LINK b.commandout TO a.commandin
```

An exitport may be linked to one or more entryports (**one-to-one** and **one-to-many** connections), and an entryport may have one or more exitports linked to it (ie. also **many-to-one** connection for a server module). The relationship between the interconnection patterns and message transactions will be discussed in the next section.

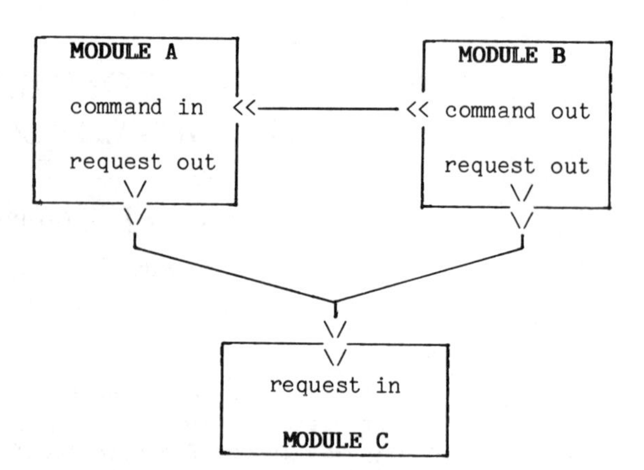

Fig 2.2 Module Instances with Linked Exitports and Entryports

Both the modules themselves and the interconnection links can be created and deleted at run time without stopping the rest of the system. This allows easy extension and modification of a control system. Since an exitport can only be linked to an entryport of the same type, this is analogous to, but more secure than, using standard hardware components and interconnecting them via plugs and sockets.

However, in process control applications configuration changes to the system must be strictly controlled and are usually the responsibility of a plant engineer. This is reflected in the system in that configuration changes are supervised by a third party, the **configuration manager** rather than by the modules themselves. The configuration manager is responsible for validating that the exit and entry port are of the same type and translating names to addresses. All the software to perform validation and checking can be contained within the configuration manager, thereby allowing simplification of the support required at each station. The configuration manager need not be centralised but could be implemented as a distributed set of modules. A few modules in a system may require the capability of sending messages to the configuration manager to modify the configuration of their exitports eg. for automatic fault recovery. In general, however, links are set up when a module is created and last for the lifetime of the module.

2.2 Communication Primitives

The CONIC Programming Language supports two kinds of message transactions for inter-process communication:

Request-reply: synchronous bidirectional,
Notify: asynchronous, unidirectional.

The same language primitives can be used for both local and remote communication. A more detailed justification of their choice is given elsewhere [Kramer 81]. A summary of the primitives is given at the end of this section in figure 2.3.

a) **Request-Reply Transactions**

This transaction supports the bidirectional "command-response" or "query-status" transactions commonly found in computer control systems. This is synchronous in that the sender is blocked until the reply is received from the responder. Similarly the responder can block waiting for requests to arrive. Request-reply transactions can only be one-to-one between a pair of modules (the semantics for multiple replies from a multi-destination request are unclear). Request-reply connections may therefore only be one-to-one or many-to-one.

Sender Task:

```
SEND command to out
    WAIT response => ... success action
    TIMEOUT period => ... timeout action
    FAIL           => ... failure action
END
```

where command is an expression of type "commandtype" and response a variable of type "responsetype". The exitport "out" would be declared as:

```
EXITPORT out: commandtype REPLY responsetype
```

Under normal circumstances the "successful send action" will be executed once the response has been received: the statement then terminates. An alternative (optional) timeout branch may be used to limit the waiting time: the "timeout action" will be executed if no response is received within the timeout period. Also an (optional) fail branch may be used to perform some "failure action" if the exitport is not linked to an entryport. For instance, permanent communication failures are a connection loss which can be indicated by unlinking the exitport. In the absence of failures the request-reply transaction will be completed exactly once. Otherwise the request message will have been received at most once.

Responder Task:

The receive-reply is used to accept a request message from an entryport and to return a response eg.

```
RECEIVE command FROM cmd;
...............
REPLY status TO cmd;
```

where command is a variable of type "commandtype" and status is an expression of type "responsetype". The entryport cmd would be declared as:

```
ENTRYPORT cmd: commandtype REPLY responstype
```

The request message received on an entryport may be from any of the exitports linked to it ie. the messages are queued and handled one at a time. As can be seen, the reply to a request is sent to the local entryport and not directly to the caller's exitport.

The receive-reply primitive may be incorporated in a select statement to enable a task to wait on a message from any of a number of potential sources (entryports). An optional guard can precede each receive in order to further define conditions upon which messages should be accepted. If the guard is false then the corresponding receive is not available for selection. If more than one message is available for selection then one is selected. Selection is made according to some 'fair' strategy to avoid starvation. A timeout can be used to limit the time spent waiting for a message. For example,

```
          WHEN command_guard
               RECEIVE command FROM cmd
               =>   .........
                    REPLY response TO cmd;
                    .........
     OR
          WHEN device_guard
               RECEIVE reading FROM device
                     REPLY signal
               =>   .........

     OR
               RECEIVE  .............

     ELSE TIMEOUT period
               =>   .........
     END;
```

b) **Notify Transactions**

The notify transaction provides uni-directional, asynchronous message sending. The sending task is not blocked and so this primitive can be used by tasks performing time-critical functions. However, in order to avoid the complexity of dynamic buffer management usually associated with asynchronous message passing and the problems invoked when no buffers are available, we have chosen to statically allocate a fixed queue of buffers, which is dimensioned at compile time, for each receive entryport eg.

```
ENTRYPORT in: statustype QUEUE 10;
```

When no more buffers are available, the oldest message in the queue is overwritten rather than causing an exception or trying to block the sender. A single buffer ("unibuffer") which is updated by successive send operations and emptied by a receive operation is particularly useful for providing the latest status information eg. to an operator display. If it is important that information is not lost, the request-reply primitives should be used.

The notify transaction can also be multi-destination which is useful for the transfer of alarms and status information. Notify connections can be one-to-one, many-to-one or one-to-many.

Transaction	**Port Declarations**	**Message Primitives**	**Transaction Pattern**	**Connection Pattern**
REQUEST-REPLY synchronous, bidirectional	EXITPORT pump:command REPLY state	SEND start TO pump WAIT response;	one-to-one	one-to-one
	ENTRYPORT cmd:command REPLY state	RECEIVE cmdmsg FROM cmd REPLY pumpstate;	one-to-one	one-to-one many-to-one
NOTIFY asynchronous, uni-directional	EXITPORT alm:alarm	SEND signal TO alm;	one-to-one one-to-many	one-to-one one-to-many
	ENTRYPORT palm:alarm	RECEIVE almmsg FROM palm	one-to-one	one-to-one many-to-one

Figure 2.3 **Characteristics of the Message Transactions**

Sender Task:

```
SEND status TO out => success action
FAIL => failure action
END;
```

where status is an expression of type "statustype" and the exitport "out" would be declared as:

```
EXITPORT out : statustype
```

Under normal circumstances the "successful send action" is executed as soon as the message has been copied from the sender. The optional "fail action" is only taken if the exitport is not linked and no message can be sent. Again permanent communications failures could result in the exitport being unlinked.

Responder Task:

The receive primitive is the same as that for the receive reply in (a) above except there is no reply part. Again, the receive can also be incorporated into a select statement.

eg. RECEIVE reading FROM in

3. IMPLEMENTATION OF THE BASIC COMMUNICATION SYSTEM

The IPC primitives described above meet the requirements for programming most real-time applications [Prince 80]. They are simple to implement and provide identical semantics for both local and remote communication. As shown below, the overheads for setting up the connections between exit and entry ports are low as are the protocol overheads of the message transactions. One of the main reasons for their simplicity is that recovery after failure requires only local decisions. There is no attempt to withdraw from a remote transaction as in Ada [Ada 80] and so very reliable (virtual circuit) communication is not essential.

The basic message transactions were kept simple and efficient so that they could be used for implementation of the system itself. The flexible configuration capabilities of CONIC mean more sophisticated remote communication primitives such as atomic transactions, or request-replies with guaranteed once only semantics can be implemented as protocol modules using the simple IPC primitives themselves. Multidestination message passing is similarly provided as an extension to the basic system. Some example extensions to the basic communication system are described in section 4.

3.1 Local Communication

The station Kernel implements both the request reply and the notify transaction protocols for IPC within a module and between modules in the same station. The station executive is responsible for setting up and clearing links between exit and entry ports.

3.1.1 Setting-up and Clearing Links

As described in section 2.1, links can be set-up and cleared dynamically, controlled by a configuration manager. All validity checks are performed by the configuration manager so as to minimise the software required in each station. Since the configuration manager provides a human interface for modifying links, it is also responsible for translating symbolic module and port names into system addresses (numbers). There is a part of the executive called the Link Manager, in each station, which receives link set up or clear messages from the configuration manager. It translates the system addresses of exit and entryports into memory addresses and modifies the relevant port data structures (see fig. 3.1).

ENGINEER	Specifies original configuration or changes in terms of symbolic names eg. LINK operator.pump TO mainpump.cmd
CONFIGURATION MANAGER MODULE	Validates engineers's configuration specifications for consistency and type compatibility. Translates symbolic names to numeric system identifiers (addresses) of the form "net.station.module.port" and sends link messages to station link manager module.
STATION LINK MANAGER MODULE	Receives link messages and translates system address into memory addresses of port data structures. Places entryport address into exitport data structure.

Fig. 3.1 Stages in Linking an Exitport to an Entryport

Existence of a link is indicated by the address of the linked entryport held in the exitport's data structure (as mentioned there is no multidestination connection at this level). There is no state information held at the entryport end except during a request-reply transaction when the sender's address is held for the reply. This allows many different exitports to be linked to a server's entryport. Clearing a link merely requires removal of the entryport address from the exitport's data structure. Internally within a station port addresses are manipulated as pointers to the relevant data structure for efficiency.

3.1.2 Local Message Transactions

a) Request Reply

When a source task executes a sendwait, the kernel obtains the address of the destination entryport from the exitport data structure. If the destination task is waiting on the entryport the request message is transferred immediately (fig 3.2) otherwise the source is queued to the entryport data structure. The source is suspended waiting for either the reply message or the optional timeout. There can be multiple messages (senders) waiting at an entryport: they are received in arrival order.

When a task executes a selective receive, the kernel builds up a list of entryports for which the guard is true. It then scans this list looking for tasks queued (messages waiting). If any, the message is transferred from the source to the receiver's message variable, otherwise the receiver is suspended waiting for a message or a timeout. If no timeout was specified it suspends indefinitely. A run-time error is caused if no timeout was specified and all guards are false as the task would be suspended forever.

When the receiving task executes a reply the message can be copied directly into the calling task's message variable as it must be waiting. In order to optimise scheduling overheads the replying task continues unless the calling task is of higher priority. There is a single copy involved in transferring both the request and the reply as this is a synchronous transaction.

```
Exitport End                              Entryport End

                                          RECEIVE
                                          ¦ blocked
                    Copy request message  ¦
SEND-WAIT           ---------------->     . process
 ¦ blocked                                . message
 ¦                                        .
 ¦                  <-----------------    REPLY
 . continue         Copy reply message    .
 .                                        .
```

Fig. 3.2 Request-Reply Transaction

b) Notify Transaction

The sender is not blocked and so the message may need to be buffered. Most implementations of an unblocked send provide buffering within the kernel, but it is difficult to dimension the size of the buffer pool. As described in 2.2, we have chosen to provide a dimensionable buffer queue at each entryport: the buffer space is incorporated in the destination module space.

When a task executes the send the kernel obtains the destination entryport from the exitport data structure. The kernel checks whether the destination task is waiting on that entryport and if so, it transfers the message directly from the source to the receiving task's message variable. If the receiving task is busy the message is copied into a buffer provided at the entryport, if necessary overwriting the oldest message queued at the entryport. The sender can continue once the message is copied out of its address space (see fig 3.3).

The kernel call to receive a queued message is identical to the selective receive described above, except that the message is transferred from the entryport buffer rather than from the sending task and there is no reply.

Exitport End Entryport End

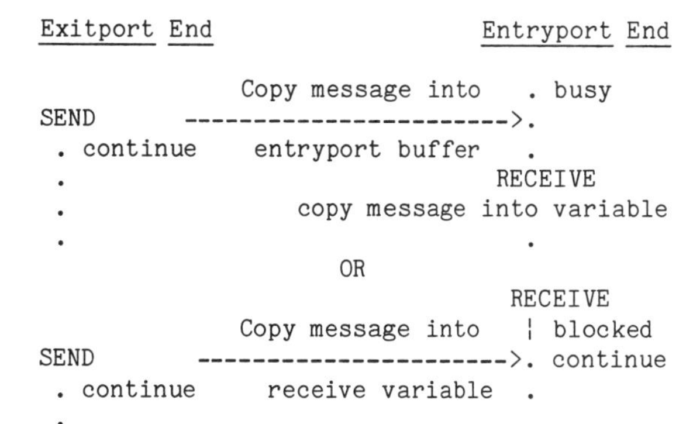

Fig 3.3 Notify Transaction

c) **Timeouts**

If a source task times-out of a sendwait, it is removed from the queue on the entryport of the destination task, even if the destination has already received the request and is busy processing it. When the destination sends a reply, there is no address recorded and so the kernel discards the reply. We rejected raising an exception in the destination to indicate a source timeout as we do not believe that there should be such asynchronous events (exceptions) in the CONIC programming language. In addition it would then be difficult to implement with identical local and remote semantics.

A timeout of a receive means no messages have arrived and so there is no clearing up required.

3.2 Remote Communication

3.2.1 Remote Exit to Entryport Links

When the Linkmanager is requested to link an exitport to a remote entryport, the exitport is actually connected to a local server entryport provided by the communication system, as shown in fig. 3.4. The exitport data structure actually holds two addresses. If the entryport to which it is linked is local, both addresses are the same but if it is remote, one holds the address of the communication system entryport and the other holds the address of the remote entryport.

The full system address of a port is a numeric identifier of the form:

subnet.station.module.port

The linkmanager maintains tables to map the system addresses of local ports into their internal representation (memory addresses).

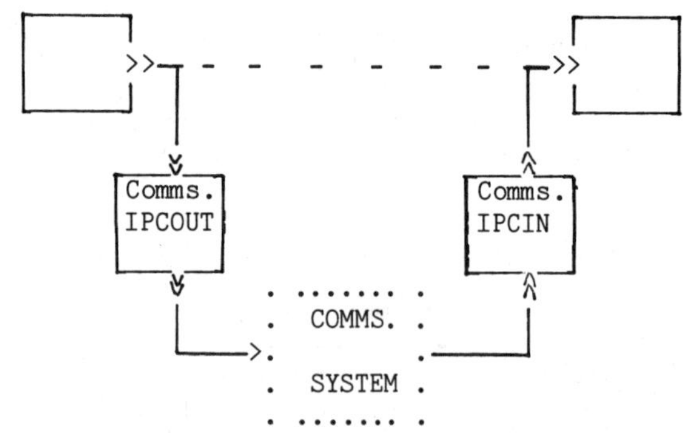

Fig. 3.4 Remote Links

When a task sends a message to a remote module, the kernel treats it as standard local IPC call and transfers the message to the communication system module. This module effectively multiplexes all outgoing exitports onto a single message stream into the network. The remote communication system demultiplexes the incoming messages onto the relevant entryports within the station. Within the communication system messages are treated as strings of bytes irrespective of their type ie. a form of type conversion has taken place. The communication system module is responsible for adding communication headers and trailers (fig. 3.5), and transferring messages to remote stations. Thus the kernel does not need to distinguish between local and remote communication.

Note that fig. 3.5 defines the frame format as seen by the IPC (or transport) layer of the communications system. It is not necessarily the format transmitted over the physical medium. **Datalength** is the length of the application message contained in **userdata**.

```
CONST  maxmsglen = 144;

TYPE
  messagetype = (request,reply,notify);
  databuffer = ARRAY[1..maxmsglen] OF byte;
  ipcframe = RECORD
                message     :messagetype;
                transaction:byte;
                timeoutval :integer;
                destination,
                source      :sysaddr;
                userdata    :databuffer;
                datalength :integer;
            END;
```

Fig. 3.5 IPC Frame Format

3.2.2 Basic Datagram Service

The communication system modules implement a simple datagram service for transferring messages between remote stations across one or more subnets. This provides a "best efforts" service in which there is no guarantee of delivery of a message to a remote destination and no automatic recovery from lost messages. This is very similar to the approach taken by Xerox in their PUP Internetwork Service [Boggs 80].

The basic communication system delivers only error-free messages. It includes error detection mechanisms and discards any messages which are corrupted. Dealing with corrupted data in a high level language with type checking is nonsensical.

This datagram service suits those applications which can accept occasional loss of data or which implement their own error recovery by means of requests and replies. Some local area networks provide error control in the hardware and so providing a similar service in software would be an unneccessary duplication for small applications consisting of a single subnet. As described in section 4, a virtual circuit service which provides retransmissions and recovery from hardware failures can be provided as an extension to the datagram service, particularly for networks consisting of interconnected subnets.

3.2.3 Remote Message Transactions

Our first implementation of remote message transactions was based on source buffering ie. the message is held at the sender's station until the destination is ready to receive it. This reduces the buffering requirement for the overall system, and facilitates discarding a request when the source timesout. However, we found that the protocol overheads of implementing source buffering were very high and so it was rejected in favour of the traditional destination buffering where a message is sent and held at the destination until received. The discarding of requests is discussed below.

a) **Remote Request Reply**

The communication module IPCOUT (fig.3.4) implements the remote IPC. When a task sends a request message to a remote destination, it actually performs a normal local request-reply to the datagram entryport of IPCOUT ie. the source is blocked waiting for the reply. IPCOUT receives the message directly into a communication buffer and builds up the communication header according to the frame format shown in fig. 3.5. It also inserts a timeout value into the header (to be used at the remote end), compensating for any time the source spent waiting for access to IPCOUT. In order to avoid further copy operations, the message frame is held in IPCOUT. A pointer to the frame is sent to the lower levels of the communication system and routing and transmission are performed directly on the frame in IPCOUT.

Once the frame has been transmitted the communication buffer can be used to send another message, as there are no acknowledgements or retransmissions at this level. We are experimenting with different versions of IPCOUT. The simplest processes a single message at a time and waits until it has been transmitted by the hardware. A more complex version provides additional buffering for multiple messages.

At the remote end, the local destination task may not be ready to receive the incoming message and so the communication module could be suspended, waiting to deliver the message. Thus a group of modules (multiple instances of IPCIN) are provided to act as surrogate senders and hold incoming messages. Fig. 3.6 shows the communication sytem modules for a station connected to a single subnet (data-link). Each free instance of IPCIN offers its services to the Datalink module by sending a pointer to its buffer in a request message ie. free modules will queue at the Datalink entryport. Datalink replies when the message has been received into the buffer of the first IPCIN instance in the queue.

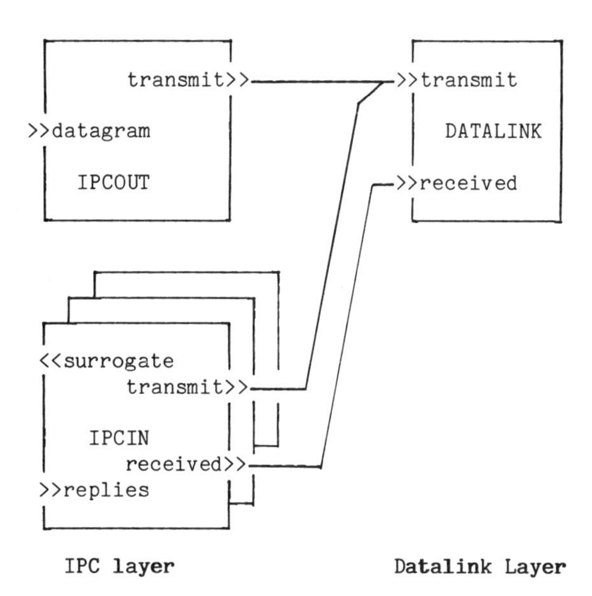

Fig. 3.6 Basic Communication System Modules in a Station

When a request message is received, the IPCIN module links its surrogate exitport to the destination task's entryport. It then performs a normal local sendwait using the timeout value provided in the message. We currently ignore transmission time as it is less than the resolution of our timeout values. IPCIN receives the reply into its communication buffer, swaps the destination and source addresses and, as was done in IPCOUT, it sends a frame pointer to the communication system for message transmission. When the reply has been transmitted that instance of IPCIN is free to deal with a new incoming message.

An incoming reply is also received by an IPCIN module. It sets up its replies entryport as though it had received the request from the source task and performs a normal reply. It is then free to deal with a new incoming message. It is feasible that by the time the reply comes the source may have timedout and sent another (possibly different) request to the destination. A transaction identifier is used to prevent an old reply being accepted by the source.

Both IPCIN and IPCOUT are implemented using the CONIC programming language and use the standard IPC primitives. However they do make use of some special kernel calls for privileged access to address information in exitport data structures and to convert between memory and system addresses. Hence they are privileged CONIC modules and should be considered as part of the station executive.

As described, we have optimised the number of message copies by passing buffer pointers and transmitting/receiving directly from/ into the buffers in IPCOUT/IPCIN. We did consider optimising the remote sendwait even further and transmitting directly from the sending task's message variable, but this would have complicated the recovery from a sendwait timeout as the transmit buffer could be in any of a number of communication modules.

b) **Remote Notify**

The kernel converts the remote notify transaction into a request with null (signal) reply, and queues the sender onto the datagram entryport of IPCOUT as above. This means the sender only continues once the notify message has been copied into a communication buffer. This was done in preference to providing a buffer queue at the entryport (as with normal notify entry-ports) as it would require two copy operations and permit possible overwriting of an older message from a different sender. The time taken to transmit a message is very small and is comparable to the normal scheduling delays in waiting for the processor. Hence the sender should not be held up for very long. However the kernel does provide a comparatively small default timeout for the request-reply. If that expires the sender continues and the notify message is discarded. We decided that discarding a notify message was preferable to blocking the sender in a real-time system, as is also the case for the local notify transaction.

IPCOUT queries the class of the exitport (notify or request reply) and sends a signal reply for the notify. This is the only difference between a notify and request reply transaction, as far as IPCOUT is concerned. At the remote end, IPCIN completes the transaction as soon as the message is copied into the destination entryport buffer. There is of course no reply required.

c) **Timeouts**

Timeouts are more difficult to deal with in remote communication because of the communication delay and the difficulty of synchronising actions. When a task timesout of a sendwait it is difficult to provide a protocol which guarantees that the request will not be received by the destination. The use of global synchronised time and a message validity time as used in Mars [Kopetz 83] would be a suitable solution, but it relies on special hardware which is not readily available or economically justifiable. Although we do include the timeout value in the message, our clocks are not synchronised to an accuracy which guarantees the message will not be received after a timeout expires.

It is however important to make sure that an old reply is not received for the wrong request. Every task has a transaction identifier which is incremented when the task timesout of a sendwait or when a correct reply is received. This identifier is sent in the request and is returned in the reply. The kernel uses it to detect and discard old replies. This is only really needed for remote communication but was easier to implement in the kernel than in the remote IPC modules.

d) **Flow Control**

All application tasks sending remote messages are held up until the message is received into the local communication system. This provides network access flow-control. Buffers are held within the communication system for a comparatively short time. The communication buffers used for outgoing messages can be freed as soon as the message is transmitted as there is no acknowledgements or retransmissions in the basic service described above.

All incoming messages are received into a communication buffer within an IPCIN module. An incoming notify message is copied straight into one of the application buffers queued on an entryport and a reply is copied straight into the waiting task. It is only incoming requests for busy tasks that are held in the IPCIN comunication buffer. Even in this case there is a limit to the time requests are held. If no timeout was specified by the calling task a system default of the maximum timeout value (10 minutes) is used. The amount of buffering (number of IPCIN modules) is a configuration option. Ultimately, if there is no free IPCIN module to handle an incoming message then it is discarded.

e) **Priority**

Some messages in a distributed control system are considered more urgent than others eg. alarms. It is possible to have multiple priority levels for messages within the system and insert messages in queues according to priority. This would cause a considerable increase in communication system complexity and result in an increase in average delays due to the additional processing overheads. Priority has therefore not been implemented in CONIC.

We did consider allowing a priority class to be associated with an exitport ie. a message takes the priority of its exitport. The same priority could be used to define the order of receiving messages from entryports in the selective receive. This was not implemented for the reasons discussed above. We do provide priority pre-emption scheduling for tasks and so we we consider it better to dedicate a high priority task to receive important messages eg. interrupt handling tasks.

3.3 Services Provided by Lower Layers

The datagram inter-station message transfer described in the previous section would correspond to the Transport Layer in the ISO Reference Model [ISO 81]. A topology of interconnected subnets needs the services of a routing (Network) layer and a Datalink layer to transfer messages across a subnet.

3.3.1 Network Layer

This layer is responsible for transferring messages across arbitrary interconnected subnets and gateways. It provides a simple internet datagram service similar to the Ethernet PUP service [Boggs 80] and is responsible for routing between subnets.

The network layer is also implemented as a CONIC module and can be omitted from a system which consists of a single subnet. Fig. 3.7 shows a communication system which includes a routing module. IPCOUT sends outgoing messages to this module which chooses the data link to which the message should be sent. IPCIN still receives all incoming messages, and those which are not for the station are sent on to the network layer.

The routing technique used is fixed routing where packets follow a path based on the shortest number of hops to a destination. The subnets are broadcast so there is no routing within a subnet, only between subnets. The routes change only as a result of a change in topology eg. failure of a subnet or gateway. Gateways are responsible for building up and updating the routing tables which indicate the next gateway to the destination subnet. The network layer addresses are of the form "Subnet.Station".

The simplest form of flow control for inter-subnet traffic is for gateways to discard messages if they run out of buffer space.

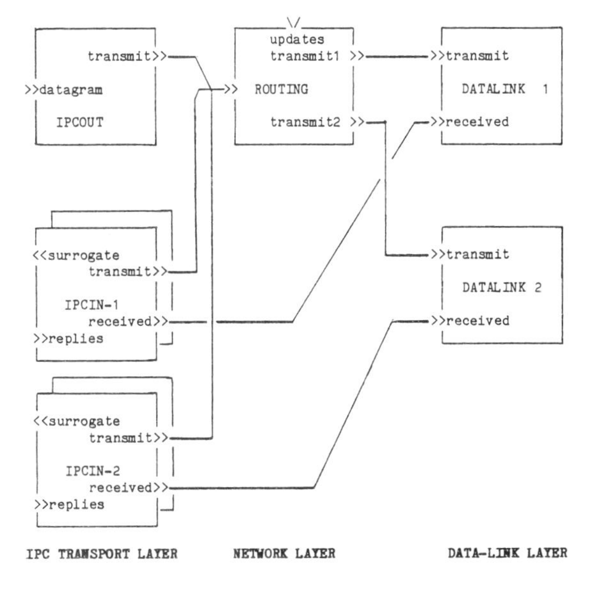

Fig. 3.7 Communication System With Routing

3.3.2 Data Link Layer and Physical Layers

This layer is responsible for transferring messages across a single subnet and for providing the hardware communication interface to the subnet. The CONIC communication system has been designed to be independent of mechanisms used to implement a subnet, and could use any suitable Local Area Network (LAN). We currently use the Omninet serial bus and will be installing a Cambridge Ring [Wilkes 79] in the near future. A suitable LAN should provide the following services:

* A mechanism for arbitrating or controlling access to the shared transmission medium connecting the stations in a subnet.
* Transfer of frames between 2 stations on a subnet.
* Broadcast of a single frame to all stations in a subnet (not provided by Cambridge Ring)
* Ability to recognise a unique station address as well as a broadcast address.
* Detection and discarding of corrupt frames.
* Bit, byte and block synchronization.

Implementation of error recovery by retransmission is not an essential requirement at this level as it can be provided on an end-to-end basis at a higher level if necessary (see section 4). Some LANs do provide such a service, particularly if the residual error rate on the transmission media is comparatively low.

4. EXTENSIONS TO BASIC SERVICE

The previous section has described the basic communication system which provides Request-Reply and Notify IPCs with the same behaviour for both local and remote communication. The basic service supports one-to-one message transactions (on one-to-one and many-to-one connections) but not one-to-many (multidestination) transactions. Also, the inter-station message transfer provides only a simple datagram type service with no guarantees of sequenced delivery or error control. Corrupted messages are simply discarded.

The philosophy taken in the CONIC system is to provide a minimum basic service and to provide additional services via CONIC software modules which can be configured into the system where required. In this section we will describe some of the enhancements to the basic communication service, namely a generalised multi-destination message transfer, reliable message transfer using virtual circuits and the interconnection of non-homogeneous computers.

4.1 Multidestination

The notify message transaction is defined as being multidestination, ie. a single send operation results in the message being delivered to multiple entryports. Within a station the simplest implemention is to make a copy of the message for each entryport. It is also comparatively simple to perform multidestination transactions across a subnet which supports broadcasting of messages, as a single message can be received by multiple stations.

The CONIC implementation of multidestination must support dynamic configuration. We allow run time extension from one-to-one connections to one-to-many without the sender being aware of the change. Also, it should be possible to transparently increase the fan-out of the multidestination exitport and to mix connections to both remote and local entryports. The obvious solution is to extend the single entryport address in the exitport data structure into a list of entryport addresses. This requires modification of the Kernel and introduces problems of dynamic space allocation. The extension is better introduced at the configuration level.

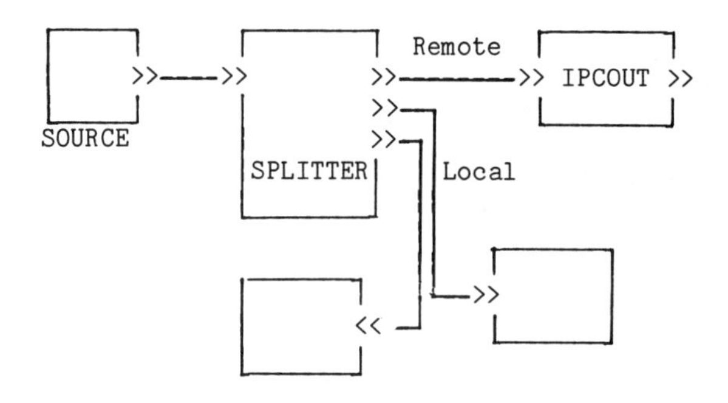

Fig. 4.1 Multidestination Splitter Module

A splitter module is introduced which directs the message to all the destinations as multiple one-to-one links (see fig 4.1). The source exitport is linked to a splitter's entryport and the splitter's exitports are linked to both local and remote entryports. The splitter module can be optimised to be merely a data structure with no code. The splitter's entryport data structure provides a linked list of pointers to its exitports. The splitter is then used by the kernel as an address mapping table. Standard configuration facilities for module creation and linking of ports can still be used.

Splitter modules can be placed in the source, intermediate or destination stations to optimise the use of broadcast transmission facilities. They can also be chained together to give additional fan-out if necessary.

4.2 Virtual Circuits

Many applications require a communication service which provides reliable, sequenced delivery with transparent error recovery eg. file transfer or application transactions which are not idempotent. These requirements can be met by a Virtual Circuit (VC) type of service. This relieves the application from consideration of communication error recovery and retransmission, although at a cost of communication protocol overheads and maintainance of virtual circuit state information at both ends. An end-to-end (transport layer) VC service will recover from errors both on the transmission medium and in any intermediate gateways.

In CONIC there is an obvious, close relationship between a link and a VC which can be exploited. One VC is provided per exit to entry port link, ie no multiplexing. The VC is set up or cleared at the same time as the exit to entry port link. VC failure can be interpreted as link failure. There is no distinction made between the VCs for request-reply and notify links: all VCs are bidirectional. As Links are always setup and cleared from the exitport end, VC set up and clearing can also be assymetric. This avoids the problems of these operations being instigated simultaneously from both ends.

The VC service provides a sequence number for each message to detect duplicates or lost messages. Sequence numbers of the last message transmitted and next message expected must be maintained at both ends as state information. A checksum is appended to each message to provide end-to-end error recovery. A received message with a failed checksum is discarded. A timer is started when the message is transmitted and cancelled when an acknowledgement is received. If the timeout expires the message is retransmitted until a retry limit is exceeded. When possible, piggy backing of acknowledgements with messages is allowed.

The initial version of the CONIC VC service was a traditional 'multi-threaded' module which provided all the VCs needed in a station. It was implemented as a CONIC module between the IPCOUT and the Network Layer. This module handled all remote VCs. When an exitport was linked to a remote entryport, the linkmanager asked the VC module to allocate a VC number and set up the VC to the remote station. The VC module had to maintain a data structure per VC which recorded current state (eg. disconnected, connecting, connected, failed, disconnecting) as well as the sequence numbers as described above.

The code was written as a finite state machine interpreter whose inputs were messages from the station or from the network. Although this is a fairly standard approach to implementing a VC service, it is rather complicated as a large part of the code deals with connecting and disconnecting VCs, and dealing with error states such as incoming disconnects for non-existent VCs. Another disadvantage of this approach is that the network layer module had to distinguish between Datagram and Virtual Circuit Services in order to route the incoming messages to the relevant entryport.

In the second version of the VC service we again tried to take advantage of the configuration facilities of CONIC by creating a module instance in both source and destination station for each VC. The state of the VC is represented implicitly by the program counter. Standard CONIC facilities for creating and deleting modules are used for setting up or clearing a VC resulting in considerable simplification of the code required. The overhead of module creation and deletion is acceptable because VCs are not likely to change frequently. In addition the Kernel automatically discards messages for non-existent modules so the handling of error states is also simplified. VC modules in the same station share code and so this approach results in much less code, although the data space required for a module is more than the data space per virtual circuit needed in the first version. Another major difference is that these VC modules reside between the application modules and the rest of the communication system (ie. VC modules are linked between the source exitport and IPCOUT and between IPCIN and the destination entryport - fig. 4.3) rather than within the communication system modules. Hence they are completely transparent to the basic communication service. The conclusion is that an extension based on existing module creation and deletion facilities results in a much cleaner and simpler approach to providing a VC service.

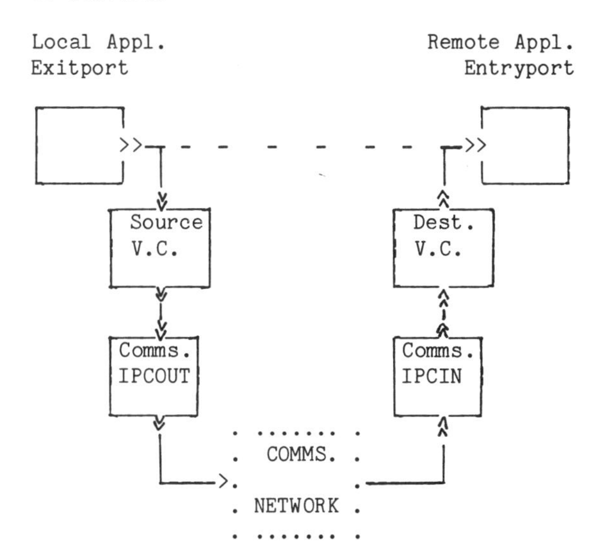

Fig. 4.3 Virtual Circuit Module Per Port

Within the communication system it is necessary to limit the length of messages in order to reduce the access time to the shared transmission medium and to reduce the buffer space required. Most process control applications only require short messages and so we chose a maximum message length of 144 bytes for the basic communication system. This allows a single message to hold a line for output to a printer and be long enough for most application messages. There is no limitation within a station on message length as no system buffers are used for local message transfer. There are a few applications in which longer messages would be more convenient and so we intend to provide an alternative Virtual Circuit Module which also fragments long messages into shorter packets and reassembles them at the destination station for delivery. This would be a configuration option.

4.3 Information Transformation

If a network consists of homogeneous computers then there is no need for the transformation of information within the communication system. If different computers communicate then the basic data units must be mapped into a standard representation for remote communication. In the CONIC architecture this is partly the responsibility of the application layer which defines message types or formats as part of application specific protocols. A compiler is responsible for mapping the typed data structures into basic data units such as integers, reals, bytes etc. recognised by the hardware.

The CONIC compiler will generate a type descriptor for global module ports. The type descriptors could be stored at both exitport and entryport so no type information is needed in messages. The copy into and out of the communication system can use the type descriptor to transform data items into a standard representation for integers, reals etc. The type descriptor could be implemented as a collection of standard procedure calls, one for each data item.

We have not yet implemented any form of information transformation as our prototype CONIC system is based on homogeneous computers. Currently a message should not contain pointers or references, but it is feasible to provide a module to dereference or "flatten" complex data structures for transfer as messages both within a station and between stations.

5. RELATIONSHIP TO ISO REFERENCE MODEL

The International Standards Organisation (ISO) have defined the Open Systems Interconnection Reference Model [ISO 81] as the basis for specifying communication standards. The networks used for distributed computer control applications do not quite correspond to the ISO view of "open" systems. Although the interconnected computer systems may support particular standards for information exchange the systems are not completely autonomous, but cooperate closely to achieve a common task. Most stations in the control network will never be accessed from outside the network. In fact for safety reasons it may be necessary to prevent any form of access which has not been predetermined. Usually the communication system, operating system and application are often rather closely integrated and are under the control of a single organisation. Introduction of new software and hardware into systems of this kind is carefully controlled. Programs are generally tested off-line and are relatively 'bug-free' by the time they are included in the system and so there is less mutual suspicion. The "users" tend to be devices or programs rather than humans. The human users are plant operators or engineers who interact with the system via well defined application programs. These application programs are usually responsible for human access control, and so it is not a function of the communication system.

We will now relate the CONIC communication system described in this paper to the 7 layers of the ISO reference model (fig 7.1).

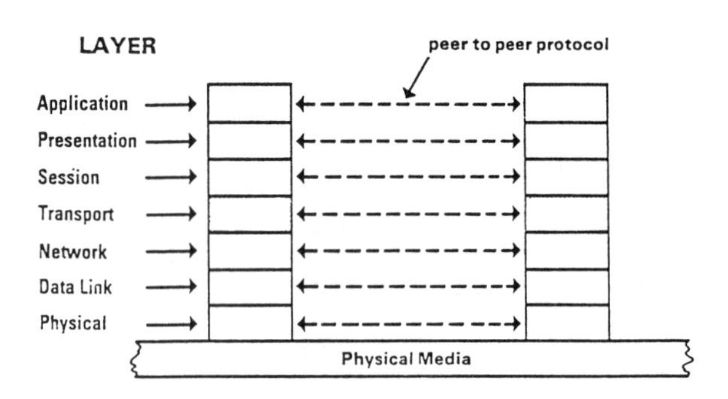

Fig. 7.1 ISO Open Systems Interconnection Reference Model

Application Layer: All the modules which perform process control, system management and operating system functions would be considered part of the application layer. The request-reply and asynchronous send are the two basic protocols available to the application layer and are the same for both local and remote communication, which is not usually the case in most current operating systems. Many different protocols can be provided (implemented as modules) using these intertask communication mechanisms. CONIC does not currently provide any particular process control protocols but does provide various operating system and management protocols eg. for remote file access or error reporting.

Presentation Layer: In the ISO model this layer is concerned with the representation of information to communicating entities in a way that preserves meaning and resolves syntax differences. In the CONIC Architecture this is partly the responsibility of the Application Layer which defines application dependent message formats and protocols. The compiler can then be responsible for ensuring standard formats for data representation. If the network consists of non-homogeneous processors it will still be necessary to transform basic data items (integers, reals etc.) into a standard representation for inter-station messages. In CONIC this would be done as a protocol module created when required.

Session Layer: In CONIC there is an association or link between an exitport and an entryport. This has some similarity to the concept of a Session in the ISO model but CONIC links can be both local and remote and are set up by a third party. In addition CONIC supports many-to-one and one-to-many connections as well as one-to-one. The ISO model currently specifies only one-to-one sessions. We really consider this layer to be part of the CONIC Operating System rather than the Communication System as it is concerned with configuration and is not concerned with actual transfer of messages.

Transport Layer: The CONIC transport layer provides end-to-end protocols as does ISO, but CONIC provides both connectionless (datagram) and connection (virtual circuit) services.

Network Layer: This Conic layer is responsible for routing messages across arbitrary interconnected subnetworks and is needed only in gateways. It uses a routing mechanism which adapts to hardware failures (transmission lines or stations) and so finds a path between any 2 subnets if one exists. It provides a simple datagram service to the transport layer. It differs from the ISO model in that it is not connection oriented and supports broadcasting in a subnet.

Data Link Layer: This transfers frames across a single subnet which could be a serial highway, loop or point-to-point line. As mentioned previously control applications will make use of LSI technology emerging for office automation applications and so we decided not to put any effort into this area but to use what could be purchased "off the shelf".

Physical Layer: This is the modem, line coupler or media access unit and is responsible for transforming information bits into transmission line signals and vice versa. Some process control applications such as coal mining and the petro-chemical industry have special requirements for intrinsically safe hardware and some may have to use armoured cables. This layer is very dependent on the Local Area Network technology used.

Relationship to Executive and Kernel

The ISO model is very unclear as to the relationship between the communication system and the executive or kernel. In CONIC the communication system is implemented as application modules and uses the Kernel provided IPC facilites. The executive in each station manages the resources for the communication modules and the network operating system manages resources on a network wide basis ie. there is no real distinction made between the communication system and the application system. The Kernel, executive, operating system hierarchy is thus orthogonal to the communication system hierarchy. Others have come to similar conclusions [Coates 82].

6. CONCLUSIONS

This paper has presented a communication system which has been designed to satisfy the application requirements, and in particular the need for local autonomy and configuration flexibility. We have presented IPC language primitives which can be used for local and remote communication. They have low protocol overheads both for connections between modules and for message transactions because they rely on local decisions for recovery from failures with a minimum backward flow of information.

The use of the CONIC language for implementation has made it very easy to provide a simple basic service and extend this by creating CONIC Protocol Modules to provide additional services wherever required. Thus the communication system can be flexibly configured to meet the requirements of a particular application. For instance, the choice of CONIC intertask communication primitives has given us the flexibility of using a variety of transport layer services. We have taken the novel approach of associating message types with application layer addresses (ie. module.port) rather than the traditional approach of using a type field within the message. For example routing updates are sent as normal application messages.

Currently the communication system is all implemented as CONIC modules and tasks. Each module is small and fairly simple. For instance, the code plus data sizes in bytes for the following modules are: IPCOUT: 896, IPCIN: 1152, DATALINK: 2038. The design has been optimised for flexibility and experimentation rather than efficiency. Despite this and the high level language implementation, the current performance for a request-reply transaction with null messages is 2.5 ms. for a local transaction and 34.5 ms. for a remote transaction. This is for LSI 11/2 processors over a 1 m.bit/s. Omninet serial highway.

The communication system may still require conversion of some of the intertask communication into procedure calls [Belanger 81] to improve performance.

We have attempted to relate our communication system to the ISO Reference Model but came to the conclusion that the ISO model does not adequately represent interprocess communication ie. the relationship between an operating system or kernel and the layers defined in the ISO model is not clear. This is particularly so in our case where the same IPC is used for both remote and local communication and is used within the communication system itself.

7. REFERENCES

[Ada 80] U.S.A. DEPARTMENT OF DEFENCE: 'Reference Manual for the ADA Programming Language: Proposed Standard Document'. July 1980.

[Belanger 81] BELANGER P., HANKINS C., JAIN N.: 'Performance Measurements of a Local Microcomputer Network'. Local Networks for Computer Communications, North-Holland 1981, pp. 181-190.

[Boggs 80] BOGGS D., SHOCH J., TAFT E., METCALFE R.: 'PUP: An Internetwork Architecture'. IEEE Trans. on Comms. Vol. 28, No. 4, April 1980, pp. 612-624.

[Coates 82] COATES K.E., 'Comparison of the Bell Labs Network Architecture and the OSI Reference Model.' Proc. COMPCON FALL 82, Sept. 1982 pp. 67 - 75.

[ISO 81] ISO/TC97/SC16/DP 7498 'Basic Reference Model of Open System Interconnection', Computer Networks Vol. 5, No. 2, April 1981, pp. 81-118.

[Kramer 81] KRAMER J., MAGEE J., SLOMAN M.: 'Intertask Communication Primitives for Distributed Computer Control Systems'. 2nd Int. Conf. Distributed Computing Systems, Paris, April 1981, pp. 404-411.

[Kramer 83] KRAMER J., MAGEE J., SLOMAN M., LISTER A.: 'CONIC: An Integrated Approach to Distributed Computer Control Systems'. IEE Proc. Part E, Vol. 130, No. 1, Jan.1983, pp. 1-10.

[Kopetz 83] KOPETZ H., LOHNERT F., MERKER W, PAUTHNER G.: 'A Message Based DCCS'. This proceedings

[Prince 80] PRINCE S., SLOMAN M.: 'Communication Requirements of a Distributed Computer Control System'. IEE Proc., Vol. 128, Part E, No. 1, Jan. 1981, pp. 21-34.

[Wilkes 79] WILKES M., WHEELER D.: 'The Cambridge Digital Communication Ring'. Proc. Local Area Communications Network Symposium, Boston, May 1979, pp. 47-61.

DISCUSSION

LaLive d'Epinay: In the CONIC work there are two different languages developed for the distributed system. Would this approach provide a good tool to manage large software systems even if they were, say, centralized?

Sloman: We build our systems using the configuration facility described since we know that system configuration is normally a very difficult process. Using the tools that we have produced we found this, in practice, to be a relatively simple task. We clearly are going to need a lot of other support tools and these are the subject of future research.

MacLeod: I would like to refer to the example which Dr Sloman gave showing the message input statements which have a guard. I would like to ask if that guard can actually involve the contents of the message, because it seems to me there is an analogy between that guard and Dr LaLive d'Epinay's firing conditions.

Sloman: No, it is essentially a logical expression and cannot involve the contents of the message.

LaLive d'Epinay: You have strong typing in the concept of the ports in the CONIC system - could you use that information in order to make type conversions from one computer system to another? Or would it mean that you would have to use different ports if you wished to transmit an integer value or, say, a real value?

Sloman: The compiler would have to generate a list of procedure calls - the compiler has the type information of the port. It could be just a record or may include arrays, reals, etc. so it might have to do some conversion from the local representation to an external standard representation to send a message. It would however increase the complexity of the information currently held in a port.

Kopetz: In the early work relating to CONIC it appeared that you were intending to use virtual circuits, it now appears that you are considering datagrams to be more significant.

Sloman: This is true, but I think it is partially because we implemented datagrams first and then found that we could live with them, but we do think there are some applications in which a virtual circuit can be very useful as well. Possibly if you knew you were going through, perhaps, a lot of intermediate nodes, involving transmitting over a point-to-point line system (which could possibly be rather noisy) then it would be easier to get rid of all the retransmission problems from the application and have standard system modules. This would mean that every application programmer would have to go through all the problems of worrying about sequence numbering and state information. Writing positive acknowledgment retransmisson protocols is not all that trivial because they start getting extremely complicated.

Kopetz: We could implement them presumably in the task of a module and, from the point-of-view of the duplication problem, would you see any difference?

Sloman: From the point-of-view of the application program you would simply create a module. As you have seen there are creation facilities and these are on a per-module basis rather than on a per-task basis. You cannot, as such, create a task and you can only create a module, so if you want something as a unit of distribution it has to be a module. So the handling would simply be in a module and the overheads would be relatively low.

Rubin: It appears to me that the CONIC structure could be useful as part of a simulation facility. Has this application been considered?

Sloman: In practice a lot of testing has been done by simulation as we have not currently got anything really to control because of the nature of the development. We found CONIC to be a very powerful simulation tool and we would like to get somebody to turn it into a general purpose distributed simulation facility. All that is really needed would be the standard things that you would need in a language like SIMULA.

Rubin: Following this I would think that the one facility which would be very useful in simulation is the capability of building sub-systems individually and stringing them together. Would CONIC be able to handle this?

Sloman: A module is effectively this. We actually consider a module to be a fairly small entity, something typically written by a single programmer and limited possibly to about six pages of listing. We have the capability of putting modules together in a way in which you do not know what is within. It is a group of modules with an interface which is probably a subset of the exit and entry ports of the individual modules. As yet we do not have this facility, but we should very soon.

Sabi-Sabi, South Africa, 1983

TASK ASSIGNMENT ACROSS SPACE AND TIME IN A DISTRIBUTED COMPUTER SYSTEM

M. A. Salichs

Departamento de Automática, E.T.S.I. Industriales, Universidad Politécnica de Madrid, Spain

Abstract. This paper studies the static allocation in a computer network of a tasks set with arbitrary and deterministic precedence constraints, in order to optimize the total processing time. A new model has been proposed, where data transmission time has been considered and parallel execution of tasks is allowed. The model is deterministic and it is used to obtain a non-preemptive scheduling. Finding an optimal assignment with this model is a NP-complete problem, so to overcome this, an heuristic algorithm has been developed. The heuristic algorithm achieves results that coincide with the optimum in most cases and gives good aproximations in the others.

Keywords. Computer control, Critical path analysis, Digital computers, Distributed computer systems, Heuristic programming, Parallel processing.

INTRODUCTION

There has not been significant progress in finding a solution to the problem of allocating in a computer network a tasks set with arbitrary and deterministic precedence constraints, in order to optimize the total processing time (T). Among all the investigations that have been made, it has been possible to achieve important results, mostly on those which work with a very simple model.

A usual simplification, used in multiprocessor systems, consists in forgetting the interprocessor communications (IPC) influence (Ramamoorthy, 1972). With this model it is possible to achieve correct results, only if the time lost in data transmission is significatively smaller than the processing time.

Some authors (Stone, 1977; Rao, 1979; Bokhari, 1981) had worked with another model where the IPC influence is introduced, but they suppose that program execution is sequential, so parallel processing is not possible, which is one of the principal interests of distributed computer systems (DCS).

We think that none of these two models is appropriate to be used in DCS. In this paper a new model is proposed, where data transmission times have been considered and parallel execution of tasks is allowed. The model is deterministic and it is used to obtain a nonpreemptive scheduling with static assignment of tasks. To find an optimal assignment with this model is a NP-complete problem. This means that it is not possible to bound the complexion time with a polynomial, so solving a medium difficulty problem would take possibly days or even months. In order to overcome this, an heuristic algorithm has been developed, which achieves results that coincide with the optim in most cases and gives good aproximations in the others.

MODEL DESCRIPTION

The computers must be all of the same type or the execution time of any task must be similar in all of them. They are linked by point to point connections and all the connections must allow the same transmission speed. In the conclusion it will be explained how to include in the model, systems with other architectures.

The program must be subdivided previously into tasks. Each task can produce a single and non divisible packet of data and each task normally needs too, some of these data packets, produced by other tasks, to be executed. This can be represented with a graph. The nodes of the graph represent the tasks and an arrow from a task "i" to a task "j" means that task "j" needs the data produced by "i" to be processed. The graph must be transitive, without loops, because the precedence constraints must be deterministic and the total processing time must be finite. A deterministic and finite loop can be always transformed into a new structure without loops. Problems with infinite loops will be studied later because this kind of problem is very common in control systems.

A transitive graph defines a precedence relationship ($<$) among tasks:

$i < j$ if there is a directed path from node "i" to node "j".

Two parameters are associated to each task:
- Processing time (Tp): the time spent by a computer to process the task
- Transmitting time (Tt): the time spent transmitting the results of the task from one computer to another.

The computers are not multiprogrammed and they are always in one of these four states:
- Processing a task "i" (Pi)
- Sending a data packet produced by task "i" (Si)
- Receiving a data packet produced by task "i" (Ri)
- Idle

With this model an optimal scheduling is sometimes obtained doing some things that could seem to be not very logical:
- Not using all available computers (Nc), e.g. Fig. 1
- Executing the same task in several computers, e.g. Fig. 2
- Sending the results of a task from one computer to another by means of a third computer, e.g. Fig. 3
- Being idle a computer, when it is possible to process some task, e.g. Fig. 3

Without change in the model it is also possible to study some conditions that are not included in this model:

(a) Sometimes it is obligatory that several tasks be executed in the same computer. This is necessary when these tasks use the same I/O physical device. Let us suppose that tasks $X_1, X_2 \dots X_n$ must be processed in the same computer. The problem can be resolved dividing each task X_i into two nonreal tasks X'_i and X''_i. The first (i.e. X'_i) with $Tp(X'_i) = Tp(X_i)$ and $Tt(X'_i) = \infty$, the second (i.e. X''_i) with $Tp(X''_i) = 0$, $Tt(X''_i) = Tt(X_i)$ and needing the results produced by X'_i. Finally it is necessary to create another non real task Y, with $Tp(Y) = 0$, that needs data issued by $X'_1, X'_2 \dots X'_n$. e.g. Fig. 4(a). With this transformation any schedule with a finite total processing time (T), allocate $X'_1, X'_2 \dots X'_n$, $X''_1, X''_2 \dots \dots X''_n$, Y in the same computer.

When $X_1 < X_2 < \dots < X_n$, then task Y is not necessary if X'_i needs the results from X'_{i-1} ($\forall i > 1$). e.g. Fig. 4(b).

(b) In the model we are describing each task originates a single data packet, but how is it possible to study problems where a task issues several data packets?. Let us suppose that tasks "2" and "3" of Fig. 5(a) need different data from task "1". It is possible to transform this graph, that is not according with the model rules, into the graph in Fig. 5(b). In the new graph task "1" has been divided into three non real tasks (i.e. 1', 1", and 1‴). All these three tasks will be allocated into the same computer because if they were allocated into different computers then the time spent in communications would be infinite.

(c) When it is necessary to repeat all the task set indefinitely and we try to obtain the maximum repetition frecuency, the best solution is not always obtained minimizing T. This problem is very common in control systems, when a control algorithm is processed continously. A quite good solution can be obtained converting the nontransitive graph, because it has an infinite loop, into a new graph where the task set is repeated several times (in most cases two or three repetitions are enough), and trying to minimize T in the new graph.

Applying condition (a) it is possible to process all repetitions of the same task in the same computer.

OPTIMAL SOLUTION

A model similar to the model here studied, but considering null trasmission times, has been investigated by several authors. They have demonstrated that this model is NP--complete. It means that it is not possible to bound with a polynomial the time spent finding an optimal solution. This model is a subclass of the model here studied, so the new model is also NP-complete. When it is necessary to resolve a NP-complete problem in a polynomially boundel time the only way is to look for an heuristic solution instead of looking for the optimal one.

The number of heuristic algorithms is infinite. It is easy to confront heuristic algorithms to know which is the best for each problem. But the best is not always good enough. The way to know if an heuristic algorithm is quite good or not, is to compare its solutions with the optimal solutions. The optimal solutions, to test the heuristic algorithms, can be preiously known in some particular cases or can be obtained with a program looking for the optimal solution in the case of very simple problems. We have chosen the second way to test the heuristic algorithm.

It has been developed a program that searches for the optimal solution of any problem according to the model. The program uses a deep-first searching method to find the optimal solution in the solutions tree. This searching method was chosen instead of the best branch and bound method, because there were memory constraints in the computer where the programs were developed.

HEURISTIC SOLUTION

The heuristic algorithm makes the assignment in two steps. In the first one, it makes a distribution of work across space, allocating the tasks each computer will process. If each task is proccssed in a differente computer, getting the maximum parallel processing, then the time spent on I. P. C. is very long. Otherwise when placing all tasks in the same computer there is no I. P. C., but that does not allow parallel processing. The best solution must be found obviously between these two extreme options. It has been previously studied that sometimes it is better to repeat the processing of the same task in several computers, than processing it in only one computer and sending its results to the computers that need these data. This is also considered in the task assignment across space algorithm.

In a second step an assignment across time is made, fixing at what time must the computing of each task begin.

Task Assignment Across Space

The task assignment across space is done in two steps:

(I) Let us begin by supposing that each task is assigned to a different computer, so there are as many computers as tasks. Then, the tasks allocated in two computers are processed in only one, decreasing the number of computers. The two task sets that are processed in the same computer are those that minimize a cost function. This process of decreasing the number of computers continues until the number of computers is equal to Nc. This algorithm does not consider that sometimes it is better not to use all available computers, as it has been studied before. But this is not very important when the number of tasks (Nt) is much greater than Nc and the precedence constraints among tasks allow enough parallel processing.

(II) After the first step, each task is allocated in one and only one computer, but it has been considered before that sometimes it is better to process the same task in two or more computers. In this step, new tasks are allocated in each computer when a costfunction is lower than a fixed value.

First, it is necessary to define the cost functions. A good distribution algorithm must try to minimize:

- The work load of each computer at any time (i. e. Overlapping cost OC)
- The time spent on IPC (i. e. Communicating cost CC).

The first objetive tries to allocate in different computers those tasks that can be processed in parallel and allocate in the same computer those that must be sequentially processed. The second objective pretends to avoid IPC. Both objectives are usually opposite.

The OC is obtained supposing that there is no IPC. These problems, i. e. without IPC, have been well studied by several authors. One of the main results obtained is that it is possible to assign two values E(i) and L(i) to each task "i", such that if task "i" is processed beginning in the time interval $[E(i), L(i)]$ then, the total processing time T is minimal.

Let us call C the computers set and P the tasks set. i. e.

$$C = \{1, 2 .... Nc\} \quad (1)$$

$$P = \{1, 2 .... Nt\} \quad (2)$$

$E(i)$ $(i \in P)$ is defined:

$$\text{if } \exists j \in P \mid j < i \text{ then: } E(i) = \max(E(j) + Tp(j)) \quad \forall j \mid j < i$$
$$\text{else: } E(i) = 0 \tag{3}$$

$L(i)$ $(i \in P)$ is defined:

$$\text{if } \forall j \in P \quad i \nless j \text{ then: } L(i) = \max(E(k) + Tp(k)) - Tp(i) \quad \forall k \in P$$
$$\text{else: } L(i) = \min(L(j) - Tp(j)) \quad \forall j \in P \mid i < j \tag{4}$$

Let us suppose that there are two tasks "i" and "j" that begin to be processed on times Tb(i) and Tb(j), each one in a different computer. e.g. Fig. 6(a). If these two tasks are processed in the same computer, trying to process them as near as possible to Tb(i) and Tb(j), then T is greater or equal than when they are processed in different computers. The difference is the overlap in the parallel processing (OVLP). e.g. Fig. 6(b).

$$OVLP(i,j) = OVLP((Tb(i), Tp(i)), (Tb(j), Tp(j))) = \tag{5}$$

$$= \begin{cases} m & \text{if } m > 0 \\ 0 & \text{if } m \leqslant 0 \end{cases}$$

$$m = \min(Tb(i)+Tp(i), Tb(j)+Tp(j)) - \max(Tb(i), Tb(j))$$

OVLP is a cost of processing two tasks in the same computer.

The beginning time of each task is not known, but it is known that, without IPC the best is to begin the processing of each task in the interval [E, L]. Using this, it is possible to define an overlap cost between two tasks (OC2) as:

$$\underset{i,j \in P}{OC2(i,j)} = OC2((E(i), L(i), Tp(i)), (E(j), L(j), Tp(j))) = \tag{6}$$

$$= \left(\int_{E(j)}^{L(j)} \int_{E(i)}^{L(i)} OVLP((t1, Tp(i)), (t2, Tp(j)))\, dt1\, dt2 \right) / (L(i)-E(i))(L(j)-E(j))$$

The funtion OVLP is not continuous, so (6) is not easy to compute fast. As we are working in an heuristic algorithm, the speed is very important, so it is better to calculate (6) fast with a not very exact method.
When there are two sets of tasks A and B, and the tasks of each set of tasks do not overlap, or the ratio between the overlap and the processing time average (Tp^*) is small, then it is defined the cost of processing the two sets of tasks in the same computer as:

$$OC(A,B) = \sum_{\forall i \in A} \sum_{\forall j \in B} OC2(i,j) \tag{7}$$

$$\text{if: } OC2(i,j)/Tp^* \leqslant \lambda_1^* \quad \forall i,j \in A \quad \forall i,j \in B \tag{8}$$

Now we are going to define CC. This cost will be different in each step of the algorithm of task assignment across space. In the first step it is necessary to consider the cost of assigning to the same computer two sets of tasks (CC_1). In the second one, it is considered the cost of adding a new task to the tasks that have been previously assigned to a computer (CC_2).

Let us suppose that there are several computers, that there is a set of tasks assigned to each computer and that each task has been assigned to one and only one computer. Let us call AS(i) the set of tasks assigned to the computer "i" and NEC(i) the set of tasks formed by the tasks that issue the data needed by task "i". The data that computer X sends to computer Y ($CMN(X \rightarrow Y)$) are:

$$CMN(X \rightarrow Y) = \{ i \mid i \in AS(X) \cap (\bigcup_{\forall j \in AS(Y)} NEC(j)) \} \tag{9}$$

If tasks allocated in X and Y are processed in only one computer, then obviously the communications between X and Y are avoided, so the communicating cost when processing in the same computer two disjoint sets of tasks A, B (A=AS(X), B=AS(Y)) is:

$$\underset{A \cap B = \emptyset}{CC_1(A,B)} = -\sum_{\forall i \in CMN(X \rightarrow Y) \cup CMN(Y \rightarrow X)} Tt(i) =$$

$$= -\sum_{\forall i \in ((B \cap (\bigcup_{\forall j \in A} NEC(j))) \cup (A \cap (\bigcup_{\forall j \in B} NEC(j))))} Tt(i) \tag{10}$$

The communicating cost of adding a new task "i" to the tasks A, previously assigned to a computer, is

$$\underset{i \notin A}{CC_2(A,i)} = \sum_{\forall j \in NEC(i) \cap \overline{(\bigcup_{\forall k \in A} NEC(k))}} Tt(j) - m \tag{11}$$

if $i \notin NEC(k)$ then $m=0$ else $m=Tt(i)$
$\forall k \in A$

This cost function considers that the IPC increases if it is necessary to obtain data that task "i" needs and that they are not needed by any task in A; and that the cost decreases if any task in A needs the data issued by "i".

The total cost of processing the sets of tasks A and B in the same computer is:

$$C_1(A,B) = OC(A,B) + CC_1(A,B) \qquad A \cap B = \emptyset \qquad (12)$$

and the cost of adding task "i" to the set of tasks A, previously asigned in a computer, is:

$$C_2(A,i) = OC(A,\{i\}) + CC_2(A,i) \qquad i \notin A \qquad (13)$$

After defining the cost functions, we are going to explain, with more detail, the different steps of the algorithm.

At the beginning of the first step, each task is assigned to a different computer, supposing there are Nt computers. The number of computers is reduced, processing the tasks previously assigned to two computers in only one, the pair of sets of tasks that are processed in one computer is the pair that minimizes C_1. But to calculate (7), the conditions of (8) are necessary. To make the set of tasks assigned to each computer verify these conditions, the next recursive function must be applied to the set of tasks that results of joining two sets of tasks.

$$Pk(A) = \begin{cases} A & \text{if } \forall i,j \in A \ (i \neq j) \Rightarrow \\ & OC2(i,j) \leqslant \lambda_1 \\ Pk(A') & \text{if } \exists\ i,j \in A \ (i \neq j) \mid \\ & OC2(i,j) > \lambda_1 \end{cases} \qquad (14)$$

where:

$$\begin{aligned} A' &= A - \{i\} - \{j\} + \{k\} \\ E(k) &= \min(E(i), E(j)) \\ L(k) &= \max(L(i), L(j)) \\ Tp(k) &= Tp(i) + Tp(j) \end{aligned} \qquad (15)$$

and

$$\lambda_i = \lambda_1^* \, Tp^* \qquad (16)$$

This function is applied only in order to calculate (7).

The second step of the algorithm assigns new tasks to each computer This step can be also defined as a recursive function F applied to the sets of tasks assigned in the first step to each computer The function F is defined as:

$$F(A) = \begin{cases} A & \text{if } \forall i \in P \mid i \notin A \Rightarrow \\ & C_2(A,i) > \lambda_2 \\ F(A') & \text{if } \exists i \in P \mid i \notin A, \\ & C_2(A,i) \leqslant \lambda_2 \end{cases} \qquad (17)$$

$A' = Pk(A \cup \{i\})$

where:

$$\lambda_2 = \lambda_2^* \, Tpc^* \qquad (18)$$

Tpc^* = processing and transmitting times average

Task Assignment Across Time

After the task assignment across space has been made in the first step of the heuristic algorithm, it is defined what tasks must process each computer. But it is not defined when the data transmissions and the tasks executions must begin nor which computer must send data to another, when this data is obtained from several computers. All of this is resolved in the second step of the heuristic algorithm. This step can be subdivided into another four steps:

(I) A new graph must be made. In this graph there is a node representing each task assigned to each computer. The nodes are clustered by computers. The data that each task needs are represented by arrows that go from the task that sends the data to the task that receives them. If a needed data packet can be issued by several tasks, processed in several computers, then it must be obtained, if possible, from the computer that needs the data; if not, it must be obtained by means of a data communication from another computer. These communications are also represented by nodes. New arrows are made from the data senders to the data receivers, passing by the nodes that represents the data communications, and if the sender is not defined the arrow from the sender is made without a beginning point. e.g. Fig. 7(a)(b).

(II) If all data senders are known, continue with step (III).

A value is assigned to each node in the

graph. This value is Tp in nodes representing tasks and it is Tc in nodes representing transmissions.

The value of all nodes affecting each computer are added. The tasks affect only one computer and the data transmissions affect the sender, if it is known, and the receiver. This addition is the work load of each computer.

Then, the nodes without data sender defined are taken, beginning with those with Tt lower and finishing with those of Tt longer, and it is assigned to them as sender the computer with lower work load and able to produce the needed data. After each assignation the work load of the computer that sends the data must be recalculated. e.g. Fig. 7(c).

(III) The parameters E and L of each node are calculated. In Fig. 7(c) those values are annexed to each node.

(IV) The time scheduling is made by executing first the nodes with lower L. When, according to this rule, a node must be executed and it is impossible to do it, then the next node must be executed. e.g. Fig. 7(d).

λ_1^* and λ_2^*

The values λ_1^* and λ_2^*, defined in (8) and (18), have not been fixed. To study the best value of these parameters fifteen problems have been randomly generated with:

Nc = 3
Nt = 10
Tp, Tt = random values with uniform distribution between 1 and 100
random precedence relations

These problems were resolved with the heuristic algorithm with five different values of λ_1^* and with $\lambda_2^* = 0$. The T obtained in each case is shown on table 1.

The averages of getting the best results with each λ_1^* value are:

$\lambda_1^* = -\infty$ 33%
$\lambda_1^* = -0.01$ 33%
$\lambda_1^* = 0$ 47%
$\lambda_1^* = 0,03$ 47%
$\lambda_1^* = \infty$ 80%

The best results are obtained with $\lambda_1^* = \infty$.

This means that the condition (8) is not verified and the function Pk, i.e. (14), is not used. There are two interpretations of these results:

(a) The function Pk is not necessary.

(b) When function Pk is not used, then (8) is not verified and the value obtained in (7) is higher, so it is necessary to multiply OC by a constant, greater than one, in (12) and (13).

We are now investigating this and it seems that the correct interpretation is the second.

With $\lambda_1^* = \infty$, as the best value of λ_1^*, the same fifteen problems were resolved again with seven different values of λ_2^*. The results are shown on table 2.

The average of getting the best results with each λ_2^* value are:

$\lambda_2^* = -0,1$ 73%
$\lambda_2^* = -0,01$ 80%
$\lambda_2^* = 0$ 80%
$\lambda_2^* = 0,02$ 87%
$\lambda_2^* = 0,1$ 80%
$\lambda_2^* = 0,4$ 60%
$\lambda_2^* = \infty$ 27%

The best results are obtained with $\lambda_2^* = 0,02$, but with $\lambda_2^* = 0$ the results are also quite good and if $\lambda_2^* = 0$ then λ_2 does not depend of Tpc^*. So the best is to use:

$$\lambda_1 = \infty \qquad \lambda_2 = 0 \tag{17}$$

HEURISTIC ALGORITHM EFFECTIVENESS

To evaluate the heuristic algorithm fifteen problems have been randomly generated with:

Nc = 2
Nt = 5
Tc, Tt = random values with uniform distribution between 1 and 10
random precedence relations

These problems have been resolved with the program that finds the optimal solution and with the heuristic algorithm. The complexity of these problems is very small, in order to be resolved optimally in a reasonable time. The results and the time spent on obtaining the results Tcpu are shown on table 3.

Tcpu is lower and more regular in the heuristic algorithm than in the optimal one. The results coincide in ten problems (67%) and in the others the heuristic results are reasonably near the optimal results.

The programs have been written in Pascal and they have been executed in a HP-1000 with a Pascal interpreter.

CONCLUSION

A new model to study how to minimize the total execution time of a set of tasks in a distributed computer system has been proposed. Although the model supposes that all computers must be linked by point to point connections. the best solution tries to minimize the IPC and this means that usually most of the physical data links are not used, so it is not necessary to implement them physically.

A program that looks for the optimal solution to any problem according to this model needs an execution time not bounded by a polynomial, so to obtain a solution in reasonable time it has been necessary to develop an heuristic algorithm. This algorithm gives quite good results but we think that it can be improved if less weight is given to the transmission cost than to the overlapping cost.

The heuristic algorithm does the assignment in two steps. In the first one is decided what tasks must execute each computer and in the second one at what time must the execution of each task begin

The first step does not consider the architecture of the communication network and it can be used with any symmetrical network (e.g. communication bus) while the second step can be adapted to these architectures easily.

REFERENCES

Adam, T.L., Chandy, K.M. and Dickson, J. R. (1974). A comparison of list schedules for parallel processing systems. Comn. ACM, Vol. 17, 685-690.

Bokhari, S.H. (1981). A shortest tree algorithm for optimal assignments across space and time in a distributed computer system. IEEE Trans. Soft. Eng., Vol. 7, No 6, 583-589.

Chu, W.W., Holloway, L.L., Lan, M.T. and Efe, K. (1980). Task allocation in distributed data processing. Computer, Nov, 57-69.

El-Dessouki, O.I. and Huan, W.H. (1980). Distributed enumeration on network computers. IEEE Trans. Comp., Vol. 31, No 1, 41-47.

Gonzalez, M.J. (1977). Deterministic processor scheduling. Comp. Surveys, Vol. 9, No 3, 173-204.

Luh, J.Y.S. (1981). Scheduling of distributed computer control systems for industrial robot. Proc. IFAC 3rd DCCS Workshop, 85-102.

Ma, P.Y.R, Lee, E.Y.S and Tsuchiya, M. (1982). A task allocation model for distributed computing systems. IEEE trans. comp., Vol. 31, No 1, 41-47.

Ramamoorthy, C.V., Chandy, K.M. and Gonzalez, M.J. (1972). Optimal scheduling strategies in a multiprocessor system. IEEE trans. comp., Vol. 21, No 2, 137-146.

Stone, H.S. (1977). Multiprocessor scheduling with the aid of network flow algorithms. IEEE. trans. soft. eng., Vol 3, No 1, 85-93.

Rao, G.S., Stone, H.S. and Hu, T.C. (1979). Assignment of tasks in a distributed processor system with limited memory. IEEE. Trans. Comp. Vol. 29, No 4, 291-299.

Problem	λ_1^* -∞	-0.01	0	0.03	∞
1	604	604	604	604	557
2	507	507	507	507	497
3	326	326	326	326	307
4	580	580	601	601	535
5	279	279	278	278	278
6	447	447	477	477	477
7	507	507	493	493	493
8	329	329	278	278	278
9	745	745	558	558	558
10	358	358	358	358	358
11	573	573	527	527	527
12	510	510	525	525	505
13	505	505	422	422	420
14	321	321	321	321	321
15	501	501	561	561	561

TABLE 1 T obtained with $\lambda_2^* = 0$

	λ_2^* -0.1	-0.01	0	0.02	0.1	0.4	∞
1	722	722	554	554	554	554	644
2	497	497	497	497	497	497	497
3	376	307	307	307	307	307	458
4	535	535	535	535	535	539	629
5	278	278	278	278	278	288	364
6	477	477	477	477	477	477	477
7	493	493	493	493	493	493	493
8	278	278	278	278	278	278	323
9	558	558	558	558	558	558	558
10	358	358	358	358	360	360	399
11	589	589	527	528	528	529	584
12	505	505	505	505	518	518	518
13	420	420	420	420	420	439	548
14	321	321	321	321	321	321	502
15	561	561	561	561	476	476	540

TABLE 2 T obtained with $\lambda_1^* = \infty$

	Optimum		Heuristic	
Problem	T	Tcpu(sg)	T	Tcpu(sg)
1	22	4.04	22	5.61
2	26	447.00	26	6.16
3	19	384.85	19	5.70
4	25	2.38	25	9.29
5	13	29.60	13	4.86
6	23	253.43	29	5.42
7	11	43.45	12	5.06
8	13	52.48	13	4.94
9	17	138.95	17	4.42
10	22	107.02	22	4.55
11	20	84.70	20	5.49
12	30	3.52	35	7.86
13	14	37.20	14	4.80
14	12	62.21	15	5.02
15	9	124.03	11	6.60

TABLE 3

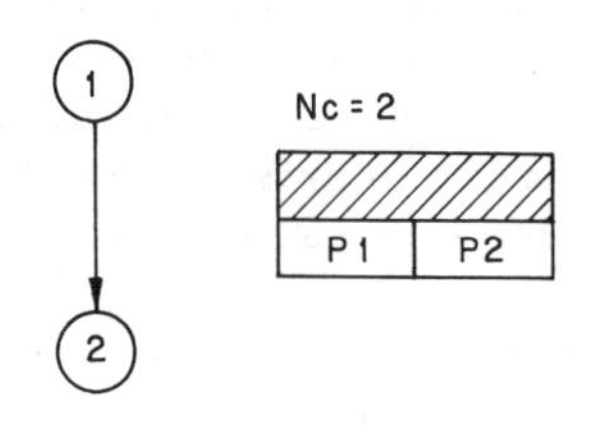

Fig. 1. It is not necessary to use all computers.

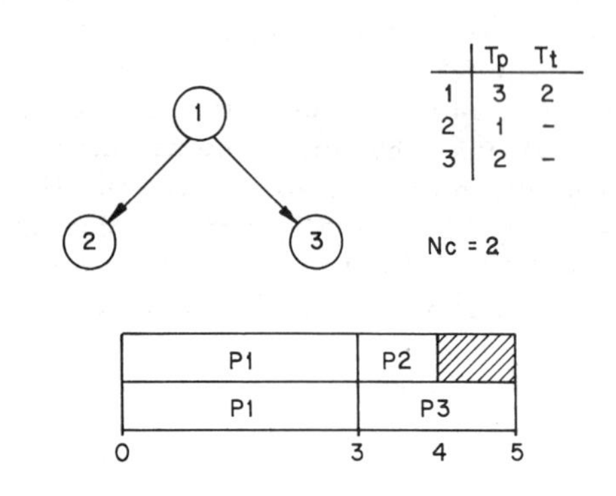

Fig. 2. The same task can be processed in several computers.

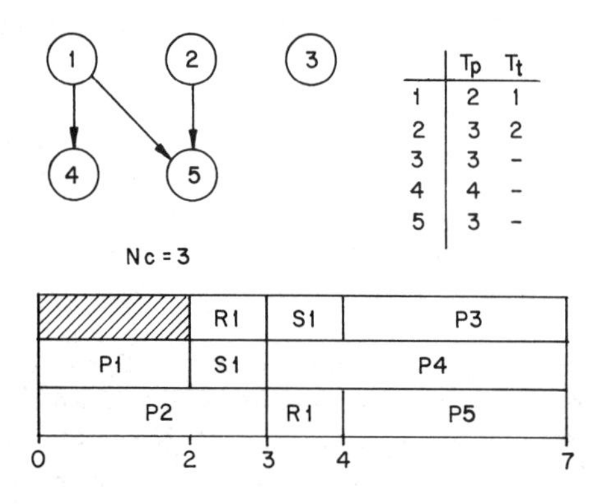

Fig. 3 A third computer is used in a data transmission.

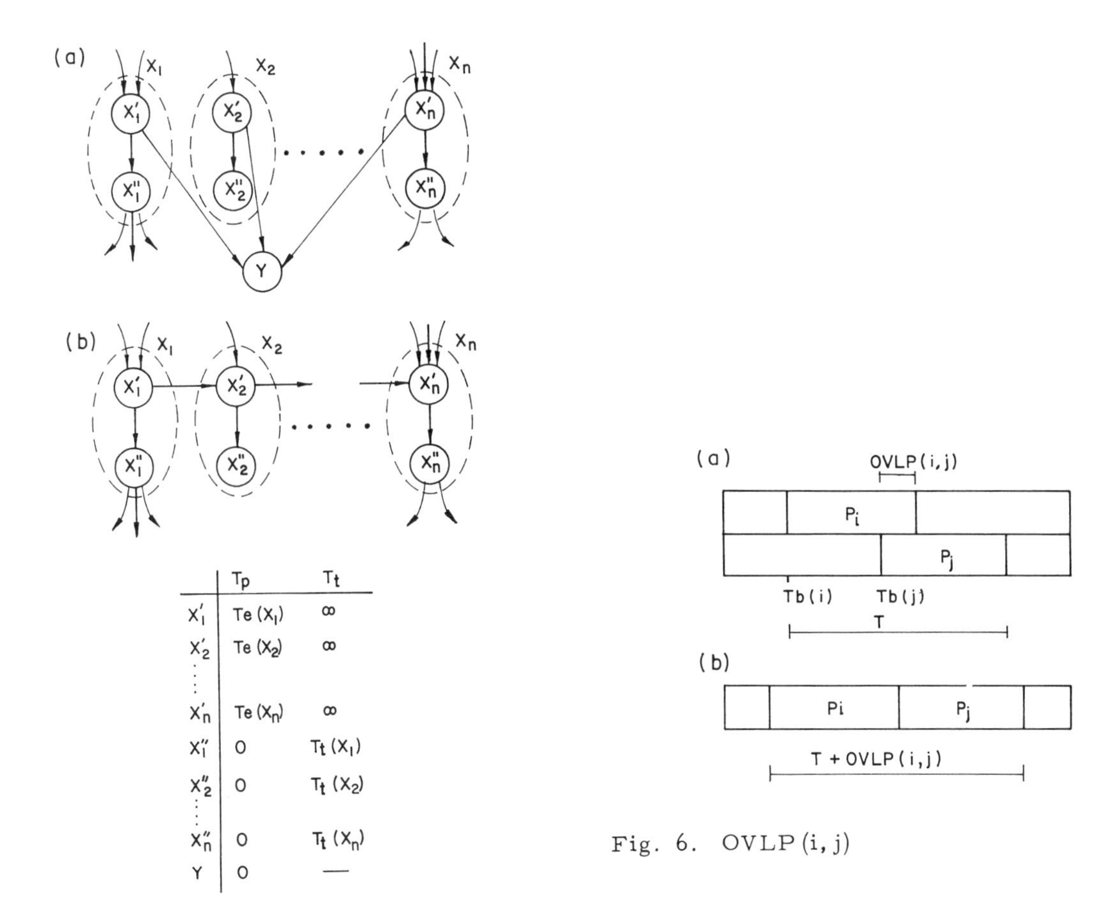

	Tp	Tt
X'_1	Te (X_1)	∞
X'_2	Te (X_2)	∞
⋮		
X'_n	Te (X_n)	∞
X''_1	0	$T_t(X_1)$
X''_2	0	$T_t(X_2)$
⋮		
X''_n	0	$T_t(X_n)$
Y	0	—

Fig. 4. Several tasks must be executed in the same computer.

Fig. 6. OVLP (i, j)

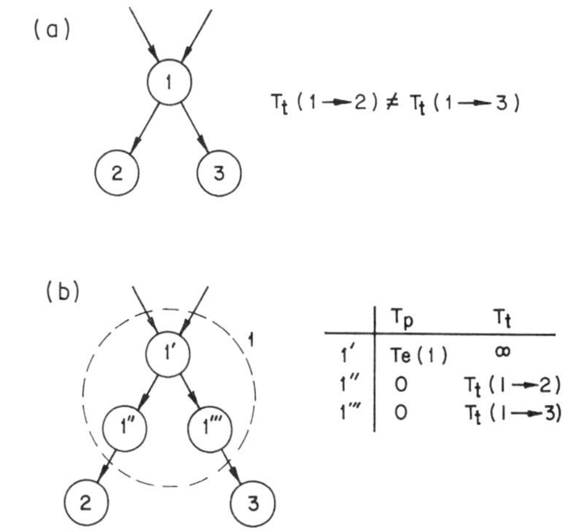

	Tp	Tt
1'	Te (1)	∞
1"	0	$T_t(1 \to 2)$
1'"	0	$T_t(1 \to 3)$

Fig. 5. A task issues several results.

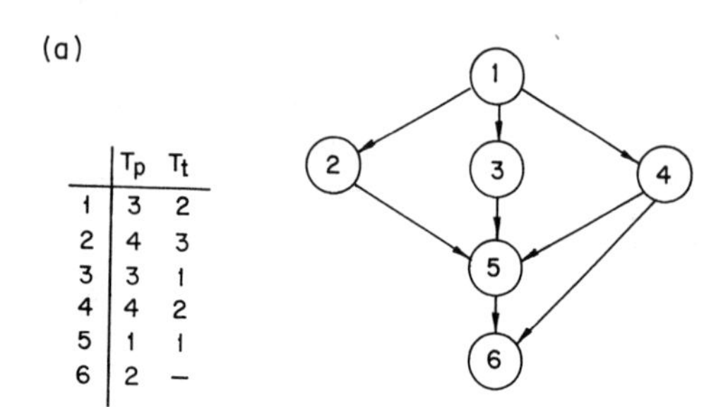

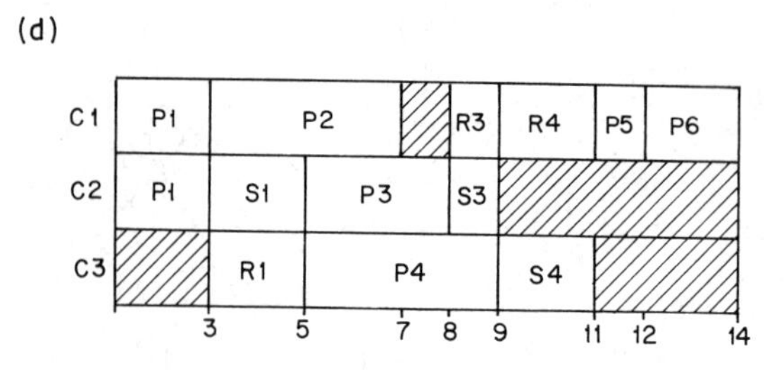

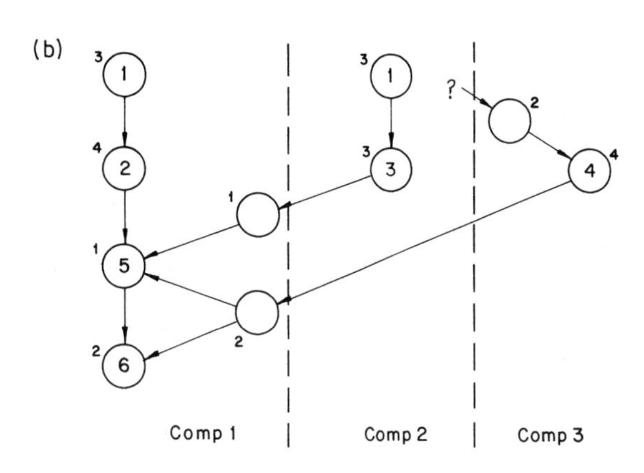

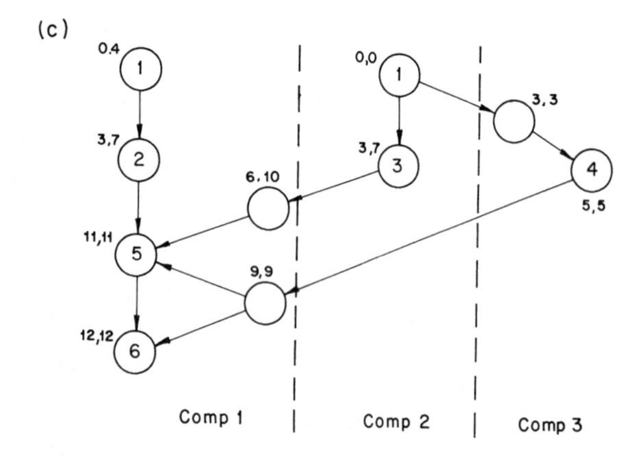

Fig. 7. Time scheduling.

DISCUSSION

Lalive d'Epinay: How much additional difficulty would it require to consider the communication system as a limited resource?

Salichs: Currently we assume that all computers are fully connected and that there is no limitation on communication. However if they are connected by, say, a packet communication system then it would be easy to transform the approach. The first allocation algorithm would be the same and it would only be necessary to make changes to the scheduling algorithm. We are currently working on this and do not feel there will be any difficulty.

Heher: Have you any thoughts about the sensitivity of the system to errors in execution times and transmission times? In a real system one does not know exactly how long a task will take to be executed since it can depend on the data.

Salichs: This is clearly a real problem of working with deterministic models. The real world is, of course, not deterministic so it is necessary to make an approximation to it. We are beginning to work with non-deterministic models but have not completed this work yet.

Rubin: It appears that the resource allocation problem is similar to project planning. In Dr Salichs' work has consideration been given to the use of standard techniques and the standard software available?

Salichs: In practice our techniques are similar to those used in PERT and there is clearly a close relationship.

LaLive d'Epinay: In real distributed systems not all the tasks can be freely moved to any computer. Many will be bound to a specific node - can this approach be introduced into the model?

Salichs: To allocate specific tasks to specific computers is relatively easy and you only have to add a new cost to the total cost function. If you have two tasks that must be computed in different computers, it is easy to put a new term in the overlapping cost. The present model can cope with this situation.

A DISTRIBUTED COMPUTER SYSTEM ON THE BASIS OF THE POOL-PROCESSOR CONCEPT

K. W. Plessmann

Department of Process-Computer Applications and Process Control, Aachen Technical University, Aachen, Federal Republic of Germany

Abstract. During the last 6 - 7 years microcomputers have been accepted in industry. Parallel to this fact, there was and is an increasing demand in distributed systems, to cope with the necessities of the running process and the reliability considerations, because the distribution of computer performance over the process is not only a question of demand of performance. For systems in use mainly on the basis of a serial data transfer the distribution is doing nothing else but providing a smaller part of a process with a processing element. This paper deals with a computer system, capable of increasing the performance by using a pool of processors to give the user the possibility to implement as many as necessary. By this, he is no longer forced to provide his system with more performance than the minimum.

Keyword. Multiprocessing, distributed processing, pool-processors, microprocessor, microcomputer, parallel processing, memory management, object-oriented programming.

INTRODUCTION

Concerning 'distributed systems' a lot of papers have been presented and published. Supported by the possibilities of the microprocessor technology the distribution of computers over the process is a step to increase the overall performance and parallel to this the reliability of the system to be controlled. Manufacturers in the field of control engineering have found it a challenging task to introduce their own concepts though the designing engineer is able to take his choice out of an increasing number of industrialized systems. Meanwhile the resulting advantages have been that great that even the mainframe vendors are offering distributed systems. At least one /1/ is structured around a serial high throughput data-bus, giving the user the possiblity to configure his systems to his demand.

Under the numerous reasons for the motivations of using distributed systems, the following are of special interest and are very often mentioned:

1. The need for more sophistication and precision in control systems, to get a maximum yield from the process.
2. The clear acceptance of microcomputer technology by a broad segment of the industry and the control engineer in particular.
3. The pressing need for clear and simpler aids for the operator.
4. The need to control the rapidly escalating costs, particularly those costs associated with the physical installation of the control systems, such as cabling and control-rooms.
5. The necessity to change the system lay-out, when modifying the control strategy or the algorithms.

While costs are not only of interest when configuring a system, the technical basis of the microcomputer technology, especially the family concept of components, has made it possible to distribute independent computer systems over the process. During the first years, mainly on the basis of 8 bit systems, it was sometimes difficult to react in a proper timeframe, but with the 16 and now the 32 bit components, we have reached a level where it is no longer of interest to discuss the possible performance.

Holding in mind that the software of a distributed system is organized so as to transfer data and to do the task, associated with the signals of the attached process, it is of some importance to state:

1. The reconfiguration of a distributed system makes it necessary to rewrite some parts of the software.
2. Splitting one system into two or more, due to the demand of the process or the lay-out, will follow in a redesign of the total program package of the station to be split.
3. The distribution of progams onto two systems, due to reliability purposes, is not only very difficult, but will influence the organisation and the program design of all other computers in the system.
4. It is often very difficult to decide under reliability aspects, how the overall program situation may influence the program design and vice versa.

These points of view are not obsolete, when the programming is done on the basis of prefabricated modules, which come together with the computer system. Today the system engineer is looking to the problems by using the interface of the given operating system. This will help to ease his implementation problems, when he is using the program modules of the vendor. When starting up a distributed system without this support, it is not only problematic to implement the software, but also will result in reliability and maintenance questions.

This paper will deal with a solution,

where a pool of processors is engaged to give the appropriate performance. The organisation of the system is done in a way in which it is possible to attach one or more processors to the pool when it is necessary and take others from the pool, in case of failure, e.g. for maintenance purposes. This will ease the application, because it is possible now to use as many processors as necessary and to change the system lay-out, under the aspects of the software system, in the case of higher or lower system demand.

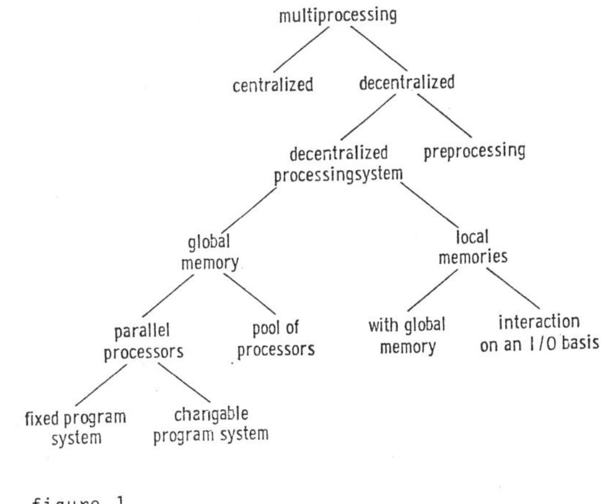

figure 1

SOME BASICS AND DEFINITIONS

A distributed computer system may be seen under the aspects of multi-computer installation, where the system components are tightly coupled, or in the sense of a loosely coupled collection of computers, controlling the peripherals, connected directly. While the first is of great interest in those installations dedicated to parallel data-stream processing, the second will be found mainly in process-type applications. Fig. 1 shows in form of a tree the different system configurations possible.

A pool system belongs to a configuration with a global memory, but is still a decentralized system. Due to its performance and construction, it may be used as a multi-computer system too.

A pool in the sense of the system under discussion constitutes a computer in the following way:

1. The overall computer system consists of a set of independently invokable processing units: the pool.
2. The active members of the pool are organized and 'attached' to a waiting task - interrupt driven or activated by the system - by a pool-management.
3. The number of processors in the pool depends only on the bus lay-out. In the system under discussion the total number is 16.
4. The processors - Pi - of the pool are logically coupled to all other components of the system, especially to the memory unit.
5. An address mapping unit - the memory manager - is engaged to translate the incoming addresses into the physically available memory space,.
6. The addressing scheme allows the definition of code and data objects.

Because of the application area of

the pool-system in the process environment, a data object may be either an operand or a message object. Both are handled in the system by the same components. While data objects are exchanged between the processors on one side and the memory or the I/O system on the other, message objects are I/O oriented only. But the interior control of the pool will check the latter to guarantee that the I/O is connected to an inter-pool transfer unit. This unit is used for the linkage of two and/or more pool-systems.

Beside this, the lay-out is done in such a way that only generally available components are used. Furthermore the concept allows the implementation of different processor-types.

Looking once again at possible applications it makes sense to organize the pool-system in two different ways. The first is used for stand alone systems, while the second, already mentioned above, creates a system of independent working pools, connected by a data transfer system. The interior control of the pool is capable to subdivide an order to transfer data into the modes pool-I/O or pool-pool and vice versa. While the structure of the pool is capable to transfer the necessary amount of data interiorly, the pool to pool exchange depends on the transfer media only.

ORGANISATION

The concept of the system is shown in fig. 2. The pool consists of 16 processors, attached to a pool manager, called SP - from: system processor. The bus system between

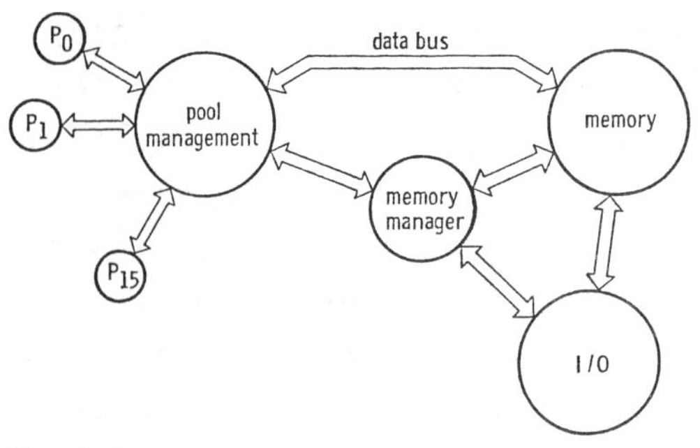

figure 2

the elements of the pool and the SP is shown in more detail in fig. 3. It is constituted of three different connections:

1. The logical addressbus ABlog. In the realisation we are using 48 address-lines, which are subdivided into two 24 busses, though modern microprocessor components, coming to the market may be applied as well as future components.
2. The databus - DB - consisting of 64 lines. DB may be subdivided into two 32 or four 16 bit subbusses. This is necessary to allow and guarantee enough flexibility. While the first installation in form of a pilot project was based on 16 bit components, the next will apply 32 bit microprocessors. Besides this it is worthwhile to hold in mind that some of these processors are sending information on their control bus concerning the actual data-width, though it makes sense to have possibilities

to use any combination of the DB.

3. The objectbus. This bus is used for all control lines necessary to drive the system but also to connect the processors of the pool to the SP during the preparation phase of the access cycle of an object. This is nothing else but a claim for an object. In response to this, the SP will lookup its tables, written during the load process and unlock the appropriate mechanisms inside the memory manager of fig. 2.
4. The memory manager - MM - is the mapping device of the logical AB to the physical. Fig. 2 shows this on the right hand side in form of a connection between the MM and the memory - SM, from: system memory - itself. In this sketch the DB is connected to the SP, while fig. 3 shows the connection in a different way - to the

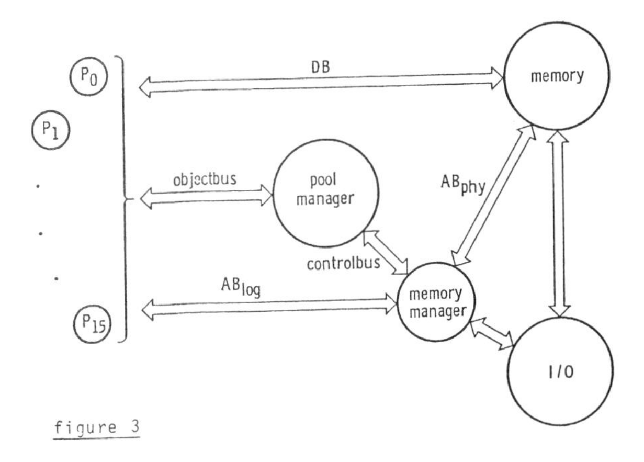

figure 3

pool. When using a bus system that was designed for the pool-processor, the last version is in use. When using a standard bus - e.g. VME /2/ or Multibus II /3/ - the first connection is engaged.

5. The I/O system is normally working in parallel to the running pool, transfering blocks of data, though it is necessary to allow access to the MM for address mapping and to the SM for the data interchange with the memory. This method implicates the basis of the object oriented addressing scheme and the access to the I/O system to ADRphys of fig. 2.

Under this consideration the pool-processor system constitutes a computer architecture that will allow and/or support the following:

1. Multiprocessing in a way that has any level of freedom, as long as 16 parallel processors are sufficient to cope with the workload of the process.
2. When less than the maximum number of processors are able to do the job, the user is free to pull as many out of the system as possible until the remaining are capable to do the processing.
3. Since the SP is managing the access to the processors of the pool and their engagement in the processing, it is possible to pull only one out of the system in case of failure or for maintenance purposes.

While 1. has to do with the overall structure of the pool-processor system, 2. and 3. are features of some importance for multicomputer applications in the industrial field. Until now it is necessary to know how many computers are involved in the solution of a problem and the coordination has to take place on the logical level of the application program: who is working together with whom. Rearranging processors in a system is possible only when the progammer knows at least how this could be done or which structures and connections could be chosen. Even this organisation was and is very difficult to do and is not only error prone, but also requires a level of understanding of the interior organisation of the computer systems and their interactions, normally not a programmer's task.

In comparison to other or similar computer architectures of this type it must be mentioned that /4 - /6 have integrated some of these features. Especially /4/ is the first industrialized version of an object oriented structure in form of a microcomputer family. Unfortunately the interior design of the main component has resulted in a lot of problems, though the performance of the whole system is not very challenging..

PROCESSORS AND SP

Fig. 4 gives an insight into the structure of one processor of the pool. Besides the different arbitrations to the bus system, a context FIFO is shown. This element is used, when a context exchange is necessary due to the demands of the present process situation and the priority of all active and waiting interrupts. The FIFO serves as an intermediate buffer for the context.

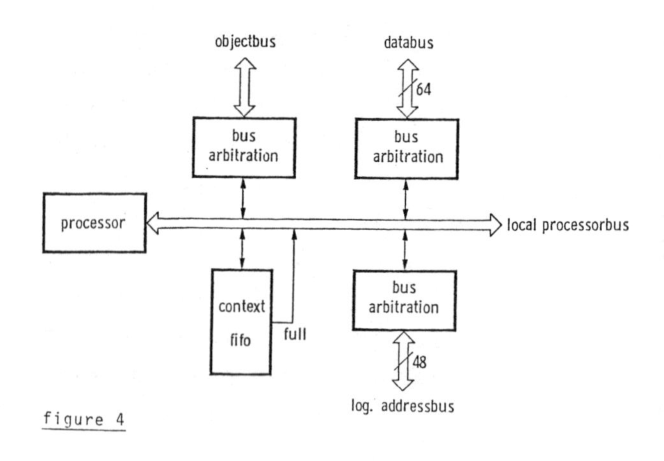

figure 4

When a rewrite has to take place, the FIFO is rotated until the context is on top and then the rewrite starts. This has the advantage:

1. not to connect the processor to the memory and
2. to decrease the context switching time, at least by a factor of 5.

When integrating another processor into the system, the arbitration has to be modified and redesigned. This is of course much easier than to redesign the total system.

Of some interest for the total structure of the system is the SP. Fig. 5 shows the architecture in some detail. The SP consists of two independent computers. The first is used to handle the objectbus, connected to the Pi's of the pool. Not shown is the connection to the MM, to control and influence its functions. By use of a FIFO the first part of the SP is linked to the I/O computer, which is connected to

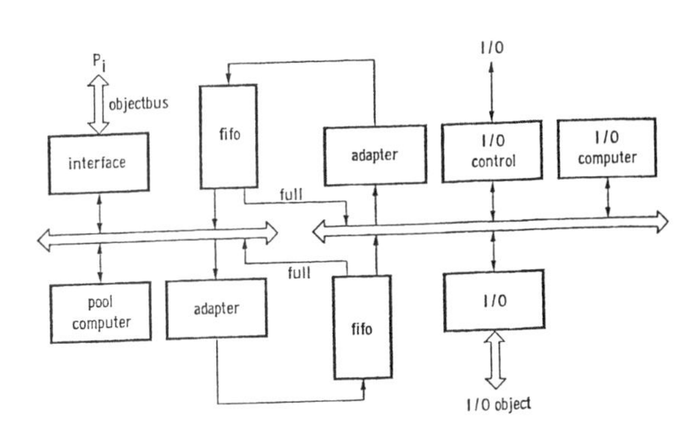

figure 5

the first in the same way.

The I/O and the pool computer - this is the first part of the SP - are microcomputers. In other words:

1. The different functions of the system - object handling, I/O functions, supervisory control, data exchange to organize the objects in the system etc. - are running in form of programs on two parallel interconnected computer systems.
2. The call for an object handling function is nothing else but the start of a procedure in one of the two.
3. The connection between the Pi's and the pool computer is organized on the basis of an I/O interchange. I/O computer is a buffer for an I/O object, which will be exchanged between the system and one or more of the peripherals.

This basis has made it possible to install all algorithms on a software basis. A modification of the system leads to a change of the driving programs. This was the most appropriate way during the pilot installation. But it is possible to duplicate the SP when it is necessary.

ADDRESSING OBJECTS

Under the aspects of a minicomputer, the addressing can be done in two ways. The first gives access to as many memory locations as the addressbus is supporting directly by its address-lines. The second is using a mapping unit because of the increasing demand in terms of memory locations and to have at least some protection mechanisms. Fig. 6 shows how this can be achieved.

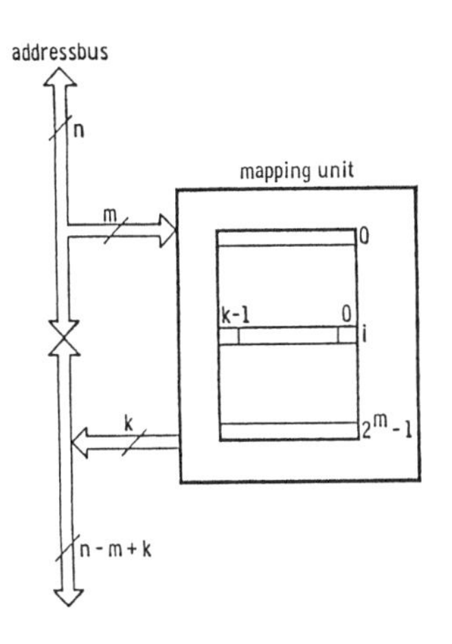

figure 6

Units of this type are also found in the component set of microcomputer families, sometimes called memory manager. The mapping unit has to calculate the incoming m address-lines into (n-m+k), where m is the number of lines, necessary to drive the map memory.

The access to a new object is initiated with a call to the SP, which in response informs the calling Pi that it has rights to access or not.

Once the object access is initiated and the calling Pi has rights to access, the addressing of the elements of the objects are addressed in the same way as it is done in today's systems. In other words: The system lay-out of fig. 2, connecting the address system to the MM by use of the SP is only for the initialisation and not for the normal access cycle. The rights to access an object are written to the MM and its tables, though it is very easy during the fetch and write cycle to state wether or not an addressing Pi is allowed to do so or not. The memory manager does not have to handle the access and the supervisor function during the address transfer to the memory.

According to fig. 7 the logical address sent by a Pi is subdivided into two parts. The first is holding

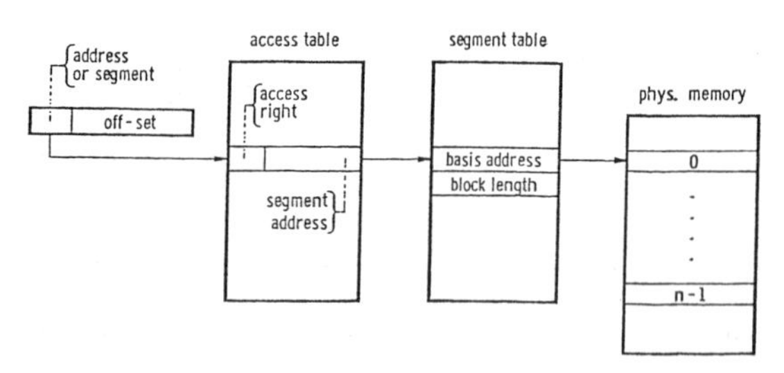

figure 7

the number or address of the appropriate access table - this is the pointer to that table - while the off-set is used to address the element inside the object under consideration.

Entering the access table is the first step to find the object in the SM. The appropriate line of the access table, entered by the first part of the processor's logical address holds the pointer to the so-called segment table. All entities in this tabel consist of two informations, the basis address of the object in the SM and the block length of the same. While the basis address is the entry point to the physical memory - in this case it is the address of the first item of an object - and the addressed item is accessed via the off-set of the logical address, the block length is used to protect all other objects. This is of great importance especially when testing the application system. Using this structure and the discussed scheme, it is possible to hide objects inside the declaring procedure and to guarantee physically the access to data in the sense of the methodology of the object oriented progamming approach.

The internal structure and the data and signal flow of the MM is shown in fig. 8. First of all the Pi is sending a signal to an encoder for the purpose of the subdivision of all actions of the MM. The output of the encoder is used to get an entry to a content addressable memory, holding the physical addresses of the object in use for this Pi. If these actions have taken place or for some of the functions in parallel, the logical address is mapped to the associative

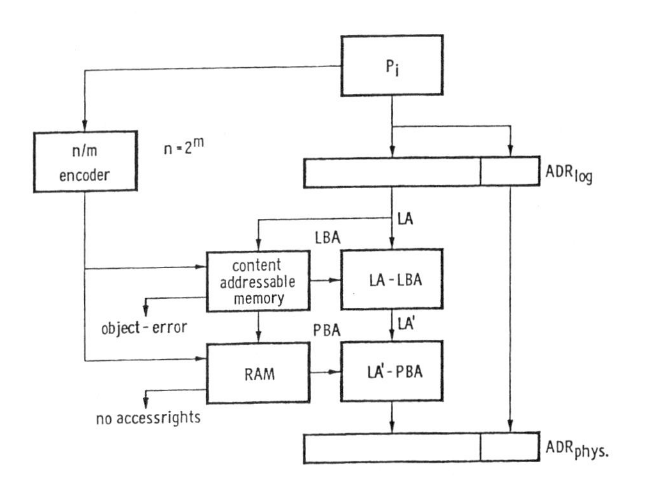

figure 8

memory, the first part is used to fetch one half of the physical address and then all other calculations, as shown in fig. 8 are taking place. Finally the basis address is read from the memory, holding the segment table and access to the addressing entry to the SM is given.

During the calculation of the address, when fetching the segment address or the access table two errors may occur:

1. There is no object of the type, or the object number called for is not connected to the program calling.
2. The calling program has no access rights, whatever the access rights may be in a system.

In both cases an error message will be sent to the SP and it will react with a system drop of the task, running on the calling Pi.

To organize a system of this kind, especially the MM, it is necessary that the SP has access to the memory manager. The data interchange is done on the basis of data and control signals, using the control bus of fig. 3. During the loading sequences of a task from the external memory into the SM the appropriate information is written into the memory of the SP and transferred to the MM, whenever they are needed. This is the case when a Pi is asking for an object during the initialisation phase, as described before.

INPUT - OUTPUT

It was already mentioned before that the system has the capability to support the I/O on the basis of a blocktransfer between the memory and the I/O system and vice versa. Using the pool processor in a technical environment, e.g. for the purpose of controlling a process, this method is not sufficient and in some instances very complicated. It is necessary to integrate an I/O system that gives the processors of the pool the possibility to read or write data circumventing the system manager and/or the MM.

Fig. 9 shows a modification, which is called the privileged I/O - this phrase intends to emphasize the special nature of the system. This form of I/O is used only for process-type applications.

Before a Pi gets rights to exchange

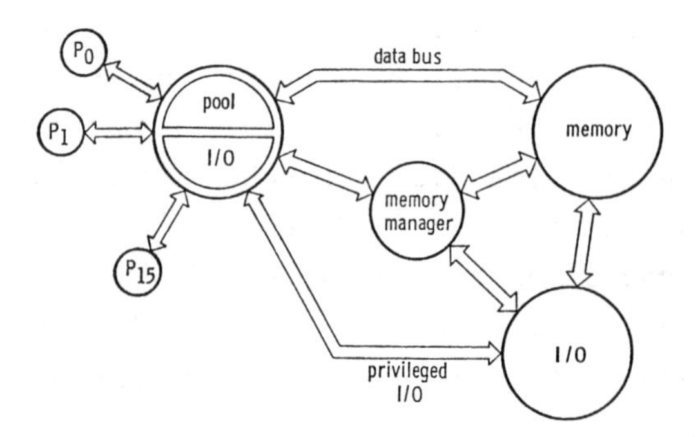

figure 9

data in a privileged mode, a call of a SP function, in the same way as the initialisation of an object to connect the I/O to the processor has to take place. The rights to access are given for a short period of time only, monitored by a timer. But during this time slice the Pi is connected directly to the I/O. To do this with the same method as the object coordination, the set of data transferred is handled as an object, having attributes of the same kind as data or code.

As can be seen from fig. 10 the I/O system itself consists of data transfer units, to support the

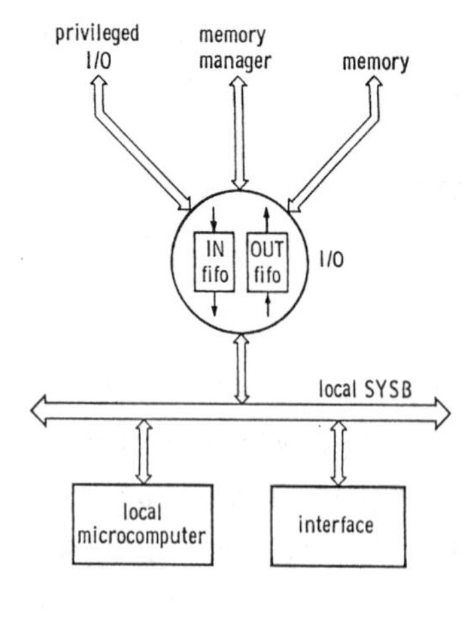

figure 10

synchronisation between the data rate of the processor system and the I/O. This is done in a very easy but powerful way using FIFO's, one for each direction. The total system is capable to handle and organize a greater number of I/O ports with this feature.

While the part of an I/O port is connected to the internal structure, the opposite side is connected to an I/O computer that is local to the I/O. This computer is responsible for the data transfer between the attached interface to a peripheral and the I/O port on the other side. It makes sense to use this structure, because the driving programs of the local I/O computer differ in the procedure to control the appropriate interface - data transfer and control/status signal interchange - and its specific behavior only. Therefore it is possible to use the same type of computer for different applications to attach peripherals, by changing the interface board and the programmable memory - i.e. PROM.

To constitute a distributed system it is only necessary to spread the I/O stations, attached to the pool-processor system over the process. This will be done in the way shown in fig. 11. The I/O module, mainly the transfer system inside, is subdivided in the following way:

1. The output-side of the FIFO for the transfer of data between the pool and the I/O system is connected to a SDLC subsystem and the local microcomputer in an I/O station has the input

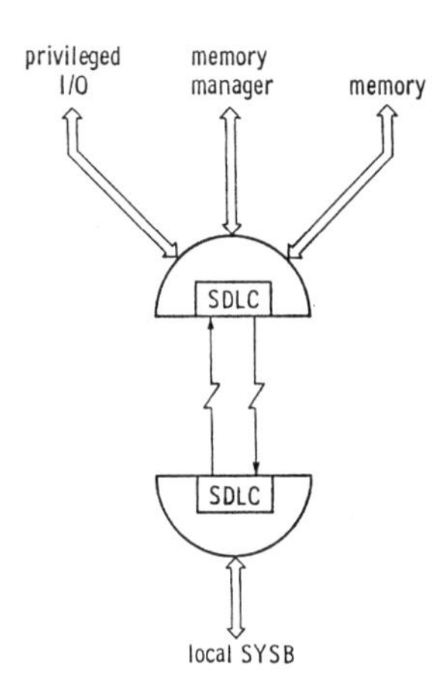

figure 11

component of the same kind.

2. The other FIFO for the transfer of data from the peripheral is connected in the same way again to SDLC, while the input-side of the I/O module is linked to the interior structure of the I/O system as discussed before.

Fig. 12 shows a block diagram of the different components involved in the

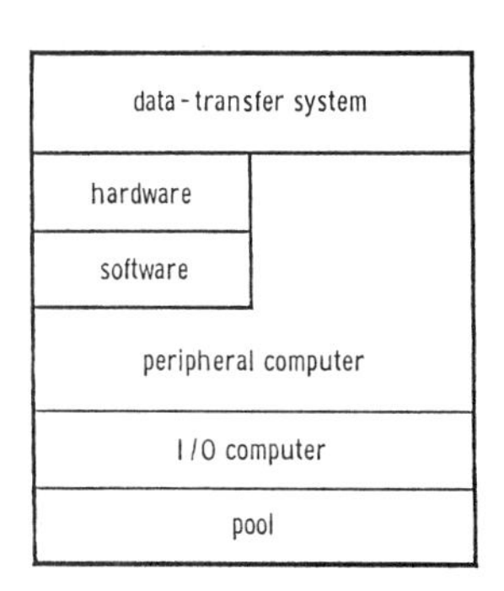

figure 12

data interchange. The hardware part of the so-called peripheral computer, together with its local I/O components - these are components of a microprocessor family - connects the pool-system in form of a star structure and allows to spread the performance of the pool all over the process.

It is obvious that a ring structure can be implemented too. In this sense the ring-master is the pool, together with the pool-manager. Under some considerations it is possible to organize the total structure in a way where one I/O system is used only. In this case preprocessing elements are attached to the ring, coordinating all the activities of the system with the SP.

The distribution of performance in this system has to be seen in the sense of a decentralized structure, where the in- and outgoing connections to the peripheral are controlled by distributed computers. This is of course a form of preprocessing, but in a way that is different from those to be found in today's systems /7/. The centralization of the main processing elements, using the pool-principle to increase the performance, while the number of elements in the pool is only a function of the demand of the process, gives the user the flexibility he needs, to face the problems of all control situations. Using standard boards, for instance on the basis of standardized bus systems, makes it necessary to integrate a standard bus instead of the local SYSB of fig. 10. The conventional Multibus /8/, would be a good choice for a system of this kind. Because of the great and still increasing number of boards, the user is free to configure the external peripheral computers in a very simple way.

SOFTWARE

The question of how to program a system of this type has to be seen under some aspects, listed below:

1. Due to the increasing performance and data throughput of microcomputer systems of today, a high level language - HLL - should be used.
2. Because of the object oriented approach it is necessary to use a HLL, supporting it or to modify an existing language in a way, to get this support.
3. Using a microprocessor family as the processing component in the pool it makes sense to use the appropriate operating system. In all cases this system may be configured to the demands of the application, to support multiprocessing and multitasking too.
4. Some of the procedures of the kernel of the operating system are already installed into the pool manager. These procedures belong to the field of the
   - message transfer between tasks,
   - handling of all activities dedicated to the different objects,
   - synchronization of tasks,
   - I/O handling, especially the privileged mode,
   - memory management and
   - exception handling.
5. The dispatching sequence is a part of the activities of the SP. Therefore there is no necessity for the user and/or the operating system to support this, in any way whatsoever.

The best choice for a programming language would probably be ADA /9, 10/. Unfortunately, the existing compilers are not very effective - this is not a severe problem - and in most cases, subsets of the full standard or the compiling process is done in two steps. The first translates the source to a PASCAL source and the PASCAL compiler has to be invoked to get a machine object. It is therefore easier to use first PASCAL and later ADA, when sufficient compilers are on the market.

To get the full performance out of the system it is necessary to extend the language at least to support the object oriented approach. This could be done for instance by implementing a data type called ABSTRACT.

REFERENCE

/1/ NN.: DOMAIN Specifications. Documentation of the APPOLO Inc., Framingham, USA

/2/ NN.: VME-System Specifications. Documentation of MOTOROLA Inc., Phoenix, USA

/3/ NN.: MULTIBUS II User Specifications and Definitions. Documentation of INTEL Inc., Santa Clara, USA

/4/ Tyner, P.: iAPX 432 General Data Processor Architecture. Reference Manual, Documentation of INTEL Corp., Portland, USA

/5/ Giloi, W.K., Gueth, R.: Das Prinzip der Datenstruktur-Architektur und seiner Realisierung im STARLET-Rechner. Informatik-Spektrum (5), Feb. 1982, pp. 21-37

/6/ Wulf, W.A. et al.: HYDRA/C.mmp. An Experimental Computer System. McGraw-Hill Book Comp., New York, 1981

/7/ Weitzman, C.: Distributed Micro /Minicomputer Systems. Prentice-Hall, Englewood Cliffs, 1980

/8/ NN.: Technical Specifications of the Multibus System. Documentation of the INTEL Corp., Santa Clara, USA

/9/ NN.: Reference Manual for the ADA Programming Language. United States Department of Defense, July 1982

/10/ Stratford-Collins, M.J.: ADA - A Programmer's Conversion Course. Ellis Horwood Ltd., Chichester, 1982

DISCUSSION

Lalive d'Epinay: At the beginning of your paper you mentioned that the pool concept allows the user to be freed from providing a system with more performance than the minimum. As I am sure you are aware there are other projects which have had similar goals and they all appear to have come to the conclusion that the additional effort required to organize the pool is at least as large as if you were to build a machine capable of the maximum performance of the the pool from the very beginning.

Plessmann: We do not feel that this is true any longer, because the number of components you need to deal with in our system is four and the component cost in our system is relatively small. Future work is being carried out in co-operation with a major supplier to produce systems which can be configurated in the way in which we propose.

LaLive d'Epinay: I would like to discuss the question of graceful degradation. My impression is that the proposed system has a lot of busses and a lot of interconnections, and I get the impression that this could cause problems from the reliability point-of-view. Did you consider the problem of graceful degradation because, if you have central elements, then graceful degradation becomes a problem. For example, can the pool manager and the memory manager gracefully degradate?

Plessmann: Clearly we did consider these problems and the central processing units will have such a problem. Therefore we planned the organization of the system in such a way that it can operate with a parallel structure - for instance, the pool may be parallel and the I/O may be parallel.

Sloman: It is interesting to note that you are pre-processing the software into PASCAL and then going out to various development systems.

Plessmann: In essence the software is being produced on a PRIME computer and then copied to two different proprietary MDS systems. The software is transferred in its high-level format and then translated by the compiler of either of the development systems.

Sloman: Our experience was that the various compilers available tend to differ and therefore make the approach somewhat problematic.

Plessmann: There is no question that this is correct, and we see the next step is to produce our own compiler which would be standard PASCAL. During the translation process we will tell the compiler what sort of object code has to be produced and what processor it is to be run on. However the approach adopted clearly leads to very rapid software development.

Rodd: In the case of a processor failure, can the system be dynamically reconfigured?

Plessmann: Although it is not discussed in the paper, there is a message transfer between the user system console and the pool processor. From this it may be found who is working and who is not, and the number of processors can be varied. This can be done dynamically.

LaLive d'Epinay: A lot of work has been done in analysing the performance of systems with various processors and memories. This clearly allows you to develop a clean model before a system is configured. Was this undertaken in the case of Professor Plessmann's system?

Plessmann: When we started the work in 1978 there was relatively little information available on similar systems and we had to attempt to design a system which was close to optimum. We did not have available any of the tools which are now available.

Kopetz: I would like to ask Professor Plessmann what happens in the case of the failure of a component within his structure? Can the system be made fault tolerant?

Plessmann: The point is that we have one task running at one time on one processor and the smallest replaceable unit would be a task. In the case of tasks operating in parallel on two different processors (one for back-up purposes) you can decide which will go on. In essence the pool processor will decide who will do which job.

LaLive d'Epinay: It appears that you cannot replace a task, you can only replace a module. Now, a task has its data somewhere in the memory and this is processed by one of the processors. So if the context, or the local data of the task, is in common memory, you should be able to continue the task if you can make sure that the processor does not destroy the task before it is killed. The problem is that you have to use a self-correcting or redundant memory because if you have to replace the memory you have to replace the data. Have you looked at this problem?

Plessmann: We have considered this problem and it does not present any difficulties.

A DISTRIBUTED SYSTEM FOR DATA COLLECTION ON A BLAST FURNACE

P. G. Stephens* and D. J. McDonald**

**Automation Department, South African Iron and Steel Corporation Ltd, P.O. Box 2, Newcastle, South Africa*
***Datalogic (Pty) Ltd, P.O. Box 5482, Weltevreden Park, 1715, South Africa*

Acknowledgement

The authors acknowledge with thanks the permission to publish this paper granted by the Management of The South African Iron and Steel Corporation (ISCOR).

Abstract

This paper describes a data collection system installed at the Number 5 Blast Furnace at Iscor's Newcastle Works. The paper reviews the operational and investigational requirements leading up to the decision to install the data collection system on an existing plant which already has a computer system installed. The system architecture is described. A powerful mini-computer is the nucleus of the Central System. Three microprocessor-based scanner systems capture real-time process data from existing transducer equipment, and transmit this data to the central system. Two microprocessor-based data entry stations are located in laboratory facilities near the blast furnace. A "workstation" computer provides the resource required for in-depth studies of plant behaviour. The workstation, data entry stations and scanners are all connected to the central system via a single multi-drop serial highway. Software for the system is based on the manufacturer's standard operating system and network control package. A standard commercial database management system, with its associated enquiry package, is used to provide the historical database facility and many of the standard management reports.

Keywords

Distributed systems; process data collection; process control; historical database; local networks.

INTRODUCTION

The ISCOR Newcastle Number 5 Blast Furnace was commissioned during December 1976. The furnace has a hearth diameter of 10.8 metres and is conveyor-charged via a bell-less top of Paul Wurth design.

In full production the furnace is capable of producing in excess of 104 000 Tons of liquid iron per month during which period appoximately 240 000 Tons of Burden materials are charged to the furnace from the stockhouse.

THE EXISTING SYSTEM

Existing Process Control Equipment

The process control equipment installed on the furnace and stockhouse is of modern design and utilises a FOX 2/30 computer System for overall Supervision, Control, Data collection and Logging. Descriptions of the Supervisory and Control aspect of the system are outside the scope of this paper; however, the Data collection and Logging system is of relevance so a brief description of each system follows.

Data collection.

Data is captured from the following plant areas

+ Stockhouse (Weighing equipment, Coke moisture gauges etc.)

+ Blast furnace top (Weighing equipment, Material gate and chute)

+ General plant area (approximately 250 points of process variables, temperature, pressure and flow)

+ Control room (Operator entered process reports etc.)

Logging.

Due to the limited amount of bulk storage available for data archiving, (approximately 500 kB) the long term storage of process data is limited to those values required for the production of averages and totals. Instantaneous values or sub-hourly averages are either reduced to hard copy form or discarded.

The following types of reports are produced:-

+ Demand Reports (These are produced on operator request)

+ Event Driven Reports (These are automatically produced as the result of a specific event, eg Furnace Charging)

+ Time Dependent Reports (These reports are produced on a scheduled basis - hourly, per-shift, daily etc. The data contained in this type of report consists of averages or totals.)

In general the data used to produce the various reports is discarded immediately upon the production of the hard copy or is limited to one backup copy in bulk memory (previous hour, previous shift, etc.)

Other reports.

To complete the survey of the reports used in the management of the furnace , mention must be made of associated data, which because of Hardware and Software limitations present at the time of system commissioning could not be incorporated into the FOX 2/30 system:-

+ Burden material analysis reports

+ Sinter Plant information

Optimisation of Furnace Production

Optimum furnace production can be described as the ability to produce the required quantity of Iron (Tons/Day) at an acceptable quality (correct metallurgical properties and temperature) at the lowest cost. The reports outlined above provide the input data for this critical management objective. Under the existing system this data is manipulated manually using certain well established criteria, the results produced being interpreted by production management to provide the necessary remedial process adjustments. The procedures necessary to obtain the results required by management are slow and labour intensive and thus prone to error. In addition, investigations which involve time-related phenomena are difficult to conduct due to problems with the synchronisation of the various data items. Despite the difficulties outlined above much progress in the optimisation of the furnace operation has been possible.

The current World-wide recession within the steel market coupled with the rapidly increasing cost of raw materials has led to a more scientific approach to iron-making practice, this in turn has led to a rapid increase in the number of process variables used in the optimisation procedure (a total of 450 required in the case of Number 5 Blast Furnace) with the process variables being scanned at a much higher frequency.

From the above discussion it can be seen that a new approach to the problems associated with the collection, archiving and processing of data was necessary to provide solutions to the problems present in the existing system while at the same time coping with the increased number of process variables.

An additional requirement is to allow the on-line automatic generation of many of the management reports which are currently produced under the existing system. These requirements led to the equipment configuration that forms the system described in this paper.

THE NEW DATA COLLECTION SYSTEM

After a careful analysis of all current and anticipated future requirements, the following parameters were established as being functionally important in the design of the new system:-

+ All data required for the generation of normal management reports to be available on-line on a fast access device.

+ Historical archiving of all the data collected by the system should be possible.

+ As investigations of a "research" nature would be undertaken from time to to time, the system must be capable of supporting this activity (if necessary down to the sensor level) without disruption of the normal data collection functions.

+ A sufficient level of data protection must exist, to allow relatively unskilled users access to the process data, without compromising the validity of the data or the continued operation of the data collection system.

+ The system should be capable of expansion by at least 100% to meet future needs.

+ The system must be capable of supporting the activities of several concurrent 'background' users.

+ As far as possible, the manufacturer's software must be used in the implementation of the system, thus reducing the maintenance requirements and the costs involved in the provision of purpose written material.

+ The hardware configuration, particularly in the scanner area, must require the minimum level of maintenance and, if possible, utilise a relay-less type of multiplexer.

From an examination of the above criteria together with a study of the plant geography the following system characteristics became evident.

+ The system should be distributed, thus minimising cabling and installation costs. This route is also seen as providing two other benifits, increased system through-put and a modular expansion capability for future development needs.

+ That a Data Base approach should be used to solve the problems associated with data access and archiving, this would also permit the use of a Query type report generator for the production of many of the management reports, thus reducing software costs and maintenance problems.

+ That a "workstation" equipped with the necessary graphics and plotting capabilities be incorporated into the system to allow the facilities required for "research" and to buffer users from the central system.

Proceeding from the outline of the background to the current project, we can summarise the essential features of the new system as follows:-

+ Process data acquisition from sensors, including conversion to engineering units, alarm checking and so on.

+ Analytical data entry, manual and automatic, from other computer based systems.

+ Historical database management.

+ Generation of management reports, scheduled and on demand.

+ Support for investigations into plant behaviour.

These features are distributed functionally and geographically across a network of seven computers.

HARDWARE

We'll look at the hardware arrangement first.

Network

The network is topologically about as simple as it could be, consisting of a single serial data link, arranged essentially in a straight line, daisy-chain fashion. Figure 1 shows this diagrammatically. In this figure we see the names of the seven computer system nodes, as well as a rough indication of the distances involved. The central and workstation systems are in a computer room located near the main blast furnace control room, and adjacent to the existing FOX 2/30 computer. The control-room scanner is in a cable marshalling and instrument rack room one floor down from the central system. The two on-furnace scanners are housed in a room built directly on the furnace structure itself, at the main operating level. Data entry station number one is in the Coal and Coke laboratory, and data entry station number two is in the Sinter plant control room. Distances are in metres.

The link is physically a full duplex serial synchronous line running over twin twisted pair lines at 56 kilobaud, using DEC's DDCMP protocol. All the line interfaces are microprocessor-based DMA type devices, with the data-link layer protocol management (DDCMP) handled by the interface, thus relieving the individual node computers of this tedious task. The interfaces used are DMP11 and DMV11 devices with integral modems, and are transformer coupled to the line, which should eliminate any ground conflict problems.

Scanners

The three scanners are almost identical, differing only in the number of actual process I/O points implemented. Each scanner is based around the LSI11/23 processor with optional floating point firmware chip. 256 kilobytes of MOS memory

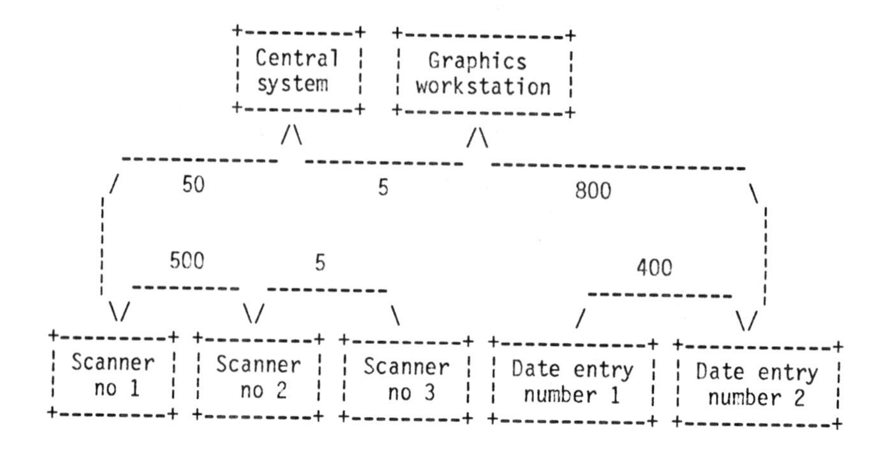

Figure 1
Network topology and line distances (metres)

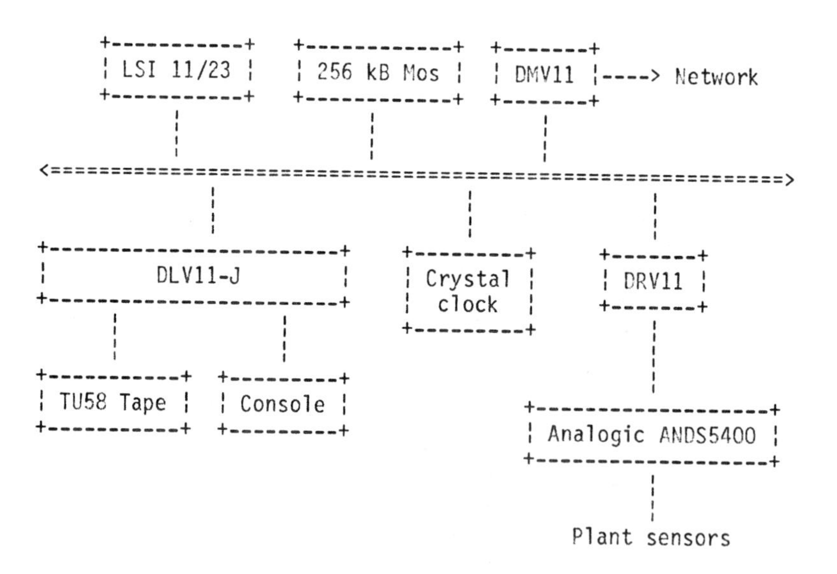

Figure 2
Scanner subsystem - hardware block diagram

with 2 hour battery backup, 4-port serial interface board, network interface, process I/O interface, and bootstrap card make up the card complement. Process I/O interface equipment is the Analogic ANDS5400 product, with the control room scanner having mainly high-level 0-10 volt inputs from existing SPEC 200 gear, and the two on-furnace scanners having primarily Type K thermocouple inputs, with a few contact inputs and a few 4-20 mA current loop inputs.

Figure 2 shows the arrangement. Each computer has a dual TU58 tape cartridge drive for local booting of diagnostics, and provision for connecting a console terminal, though no console is present during normal operation.

The scanners are housed individually in sealed steel cabinets, the on-furnace scanners being fitted with closed cicuit heat exchangers and purged with air.

Data Entry Stations

The two data entry stations are identical, and very similar to the scanner systems. The main differences here are

+ Console VDU terminal permanently installed
+ System is mounted in a small cabinet slung under a desk, much like the drawers on a standard office desk
+ No process I/O
+ No battery backup

It is expected that the Data entry system number one will have a small (10-20Mbyte) winchester disk added soon.

Graphics Work Station

This system (figure 3) is built around a PDP 11/24. It includes 256 kilobytes MOS memory, a 10.4 Mbyte removable cartridge disk, 180 cps console printer, VDU terminal, TEKTRONIX 4012 graphics VDU, and TEKTRONIX 4662 plotter. Network interface is a DMP11.

Central System

The nucleus of the central system (fig. 4) is a PDP 11/44 with FP11 floating point unit, fitted with 512 kilobytes of MOS memory with battery backup. Two 10.4 Mbyte removable cartridge disks, two 160 Mbyte (logically four 67 Mbyte) winchester disks and two half-inch reel-to-reel tape drives make up the bulk memory. The two winchester disks and two tape drives are intended to operate in a 'cold standby' mode. A 180 cps console printer and three VDU terminals are provided for online program development. In the main control room, two colour semi-graphics units and a 120 cps printer provide the operator interface to the system. Network interface is a DMP11. Simple asynchronous links are provided to the FOX 2/30 computer and an X-ray spectrograph.

Power Supply

A 15 kVA motor generator set supplies 50 Hz power to the central, workstation, and scanner systems.

SOFTWARE

Software Summary

Turning now to the software, we can start by listing in rather more detail the functions which are to be supported.

+ Acquisition of process data from sensors, including sensor validity checks (eg open thermocouple), conversion to engineering units, linearisation, filtering, alarm checking against several limits and gathering of statistics such as maximum, minimum, average and so on.
+ Formatting of three-minute 'snap-shots' of the plant for storage in the historical database.
+ Performing of 'investigational scanning' of small subsets of the plant sensors, for special purpose investigations
+ Acquisition of charging data from the FOX 2/30 computer for storage in the historical database.
+ Acquisition of analysis data from the Philips X-ray spectrograph for storage in the historical database.
+ Acquisition of analysis and operational data from personnel in three distinct geographical areas.
+ Provision of process displays, with real time update, to operators in main control room.
+ Production of complex performance parameters, with monitoring against alarm levels and storage in historical database.
+ Management of 'archive' data, including automatic spooling to tape, deletion of outdated information from online database, and subsequent retrieval of archived data for investigational purposes.
+ Production of regular management reports of plant performance and production - monthly, daily, shift etc.

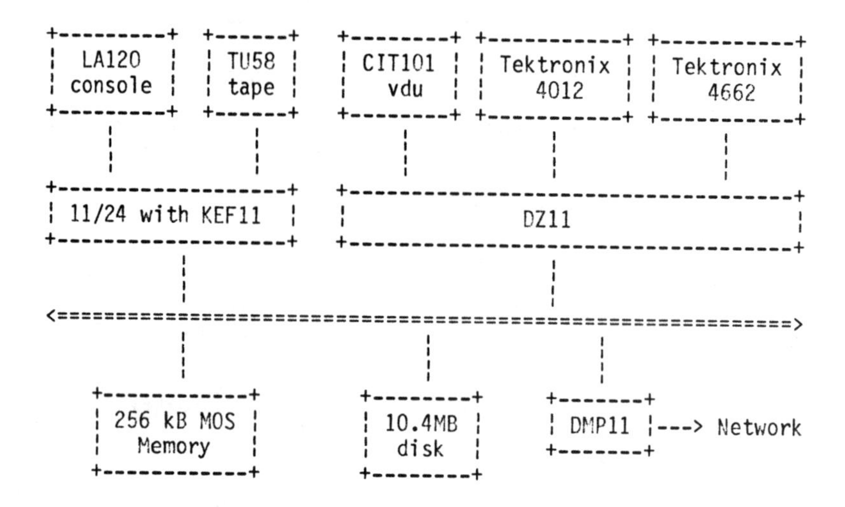

Figure 3
Graphics workstation block diagram

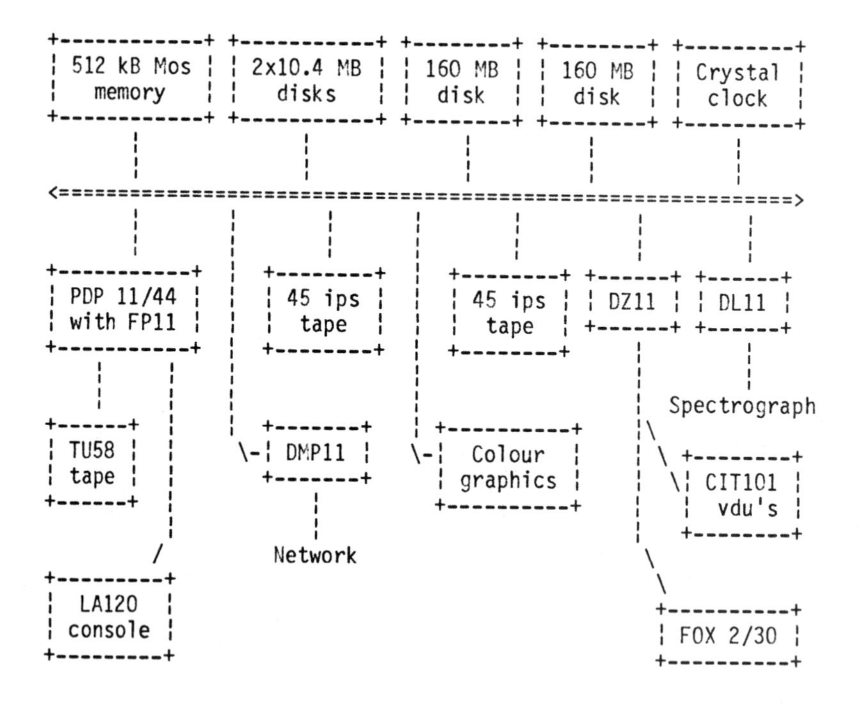

Figure 4
Central System block diagram

+ Production of ad hoc reports on a once off basis with wide flexibility in selection of report contents.

+ Comprehensive support for detailed statistical analysis of plant behaviour - modelling.

+ Support for on-line program development with minimal impact on 'foreground' activities.

From these requirements, together with the geographical constraints, we drew up a data flow diagram, figure 2.5. This diagram concentrates on data entities, which may be items of hardware eg a printer, or a file or table in memory or on disk, and the directions of flow of data (the broad arrows) and control (the single lines). Interposed between the data entities are blocks representing coherent or closely related processing steps. A block does not in general correspond to a task or program, though it may do so in a few cases. Rather, the 'inner' blocks are processing subsystems, with clearly defined interfaces and functions. Having said that, it is true that the structure of this data flow diagram does in practice tend to have an influence on the subsequent decomposition into tasks or programs. It is obviously very important that this data flow structure be carefully thought through, as it forms the foundation for the subsequent refinement of the software design.

As this workshop is all about distributed computer systems, it is appropriate to concentrate on those parts of the software which are in some sense distributed. This will have to excuse what may seem at times to be 'glossing over' otherwise important parts of the software stucture.

Operating System Software

All the systems are based on DEC's RSX11 V4. Systems with disks naturally use RSX11-M, while those without disks use the stand-alone version, RSX11-S. Network software is DECNET-11, which provides a rich set of task to task, task to remote file, and operator to remote file capabilities, both point to point and routed. Programming languages are Fortran 77 and PDP11 macro assembler. The DBMS used is TOTAL, by CINCOM Systems. TOTAL was chosen because it has adequate functionality, including a friendly query package, is a mature product on PDP11, is acceptably priced and has local representation.

The graphics workstation has, in addition to these, BASIC Plus 2, which is a powerful BASIC language compiler, and PLOT-10, the Tektronix graphics support package.

Scanners

Scanner software falls clearly into two parts, the routine data acquisistion and everything related to it, and the investigational scanning.

The primary scan task itself is a table driven generalised scanner which has no specific application data such as ADC gains, point names and addresses and so on. It is linked to a system global area (RSX resident common) where all the tables are located. Parameterisation of the task to a particular set of process points requires the production of 'point definition' statements in a text file in the general form <parameter-name>=<value>; in a relatively free format fashion. These definitions are translated, assembled, and task-built to produce the scan tables. This approach means that all three scanner computers have software which is essentially identical except for the point definition tables.

Alarm messages are printed by sending a data packet describing the message to a separate task, so that the scan task itself does not get tangled up in terminal I/O. In this particular project, the scan task and the alarm printing task are in different computers, the data packets being sent across the network link. There are thus three scan tasks capable of generating alarm messages, and one alarm printing task in the central system.

Process current values are held in a section of the system global area, where they are accessible to the task responsible for the gathering of the three-minute 'snapshot' frames. This task picks up the current value, minimum, maximum, and average for the preceding period, as they have been generated by the scan task, copies them to a buffer in allocatable memory (RSX create region) and clears the statistics counters. It then attempts to transmit this buffer to the central system. If the line is down for any reason, these buffers will accumulate in memory until the line is up again, or until allocatable memory is exhausted. If memory is full, as determined by failure of the attempt to create a new buffer, the oldest buffer currently in memory will be overwritten. When the line does eventually recover, all queued buffers will be transmitted to the central system as fast as they can be accepted.

The investigational scanning task is much simpler than the main process scanning task. It receives from the central system a table defining a number of point addresses, the number of readings to take on each, and the rate at which to take these readings. A grand total of 1000 readings on all points together has been identified as adequate for this function. The buffer containing the

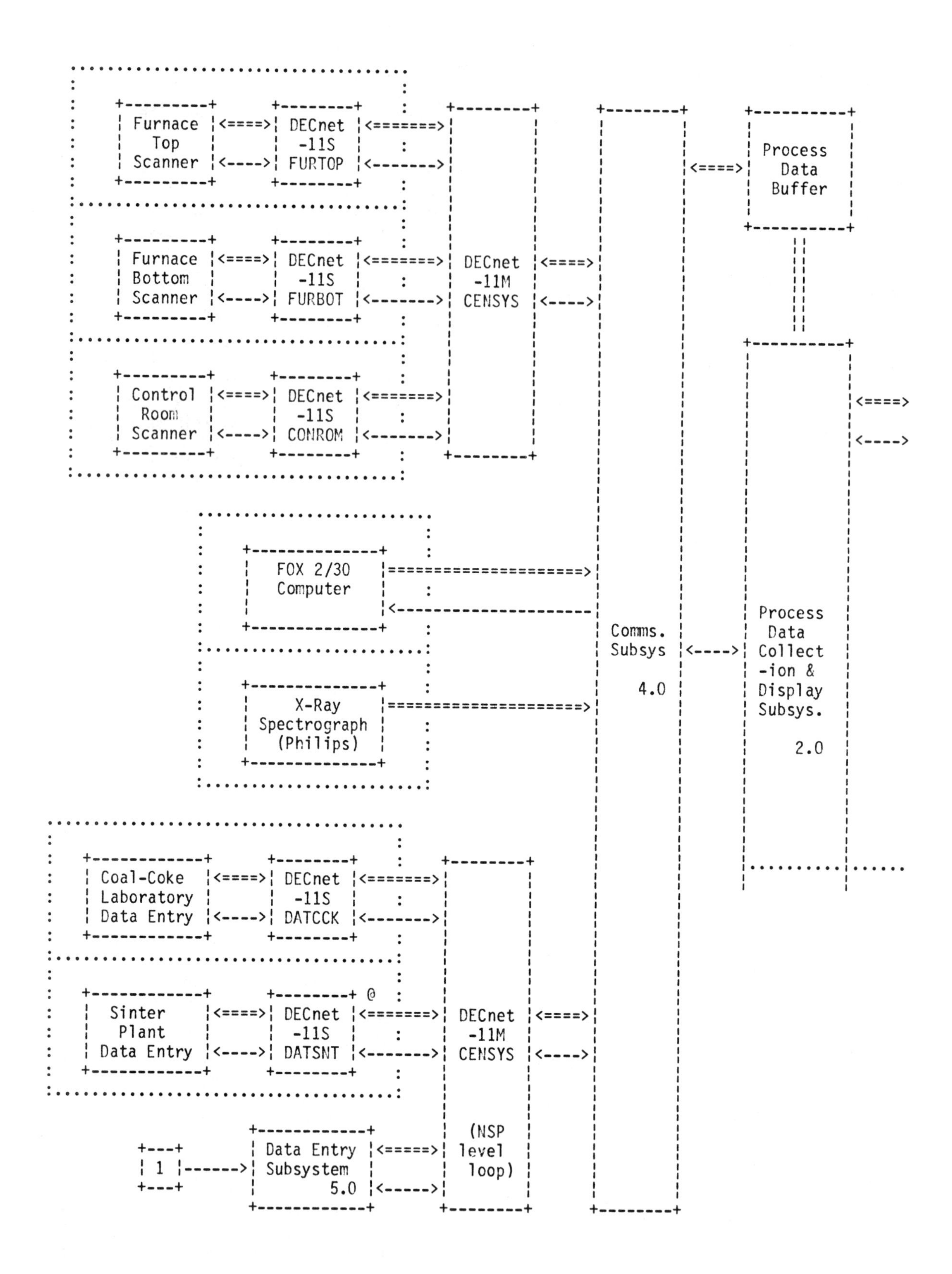

Figure 5 (a)
Overall Data Flow Diagram

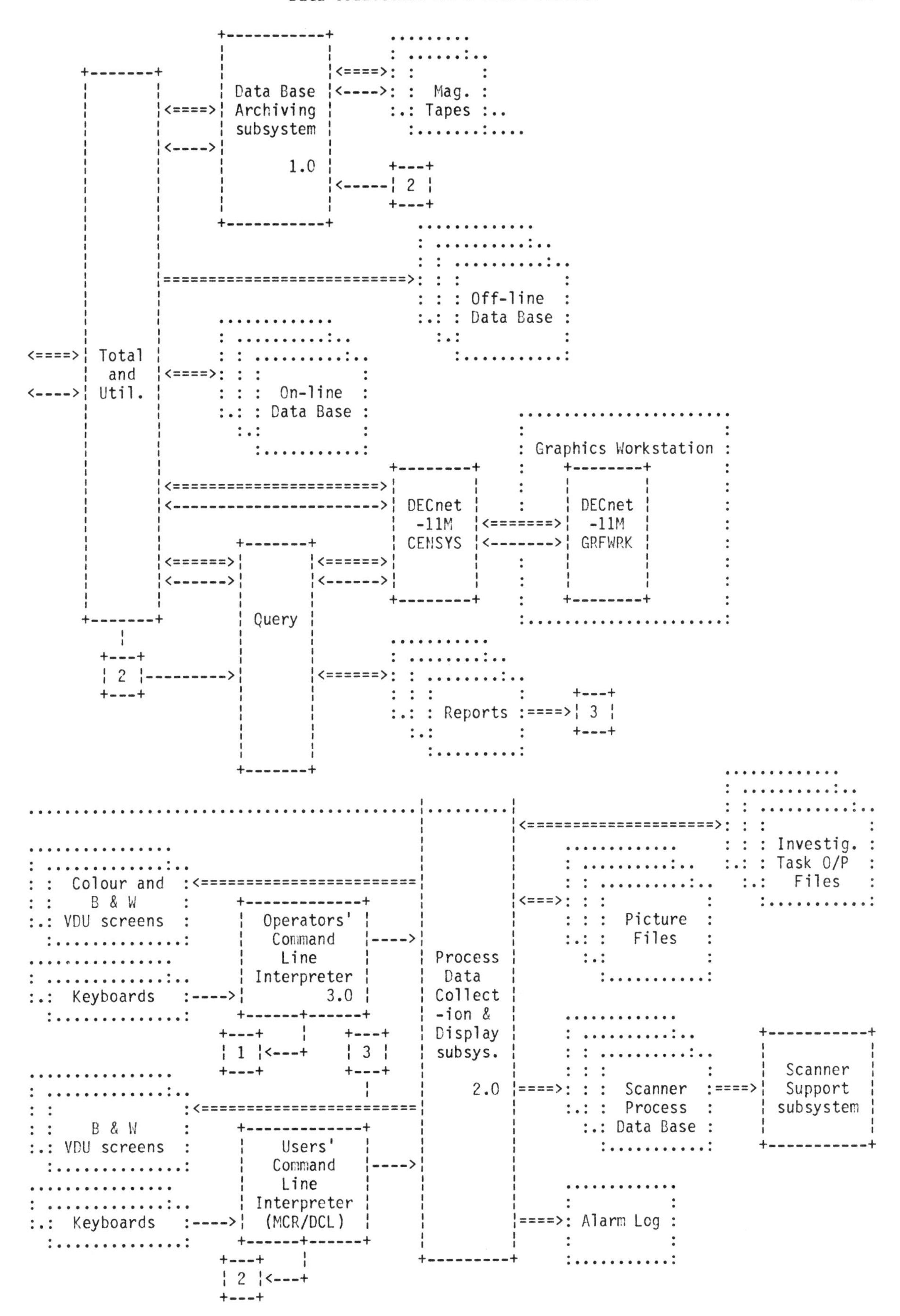

Figure 5 (b)
Overall Data Flow Diagram

results of the scan is dispatched back to the central system, where it is stored in a file for further processing by the requestor.

A common network handling task is shared by both scanning systems. This has the benefit that the complexities of network line management are confined to one task, and the remaining tasks can use simple send data or pass region directives.

Data Entry Systems

The Data Entry systems are used for the entry, by operators and laboratory staff, of operational and analysis data for incorporation into the historical database, from where it is used in the production of management reports, and analysis of plant behaviour by users at the graphics work station.

An interesting aspect of the data entry systems is that the facility is required at three geographical locations, two remote from the central system, and one at a terminal which is itself connected to the central system, while preserving a uniformity of operator interaction protocol.

The data entry function itself involves presenting the operator with a menu of forms from which to choose. The menu, and the individual forms' contents, are stored in the central system main database, with the objective of making sure that only one copy of the forms' definition exists in the system, to simplify maintenance. Once a form has been selected, the form contents i.e. form title, column headings, item name, description, format of data, validity check codes etc., are fetched from the central database, and the display put up. The operator will then key in his data. This will be subjected to the checks as defined, and appropriate interaction with the operator takes place for any incorrect or invalid data. Some examples of the checks used are

+ Numeric, possibly with range checks

+ Alphanumeric, possibly with a limited vocabulary

+ Consistent time data relative to other entries in the historical database

+ Correct sequence of entry of different operational reports

In some of these cases, the values against which checks must be performed will be in the central database, so these must be fetched to allow the check to be made.

In a manner analogous to the case of the scanners, the processing aspects are separated from the line handling aspects. This provides a convenient way of achieving the requirement that identical facilities be available at remote nodes and local terminals, while at the same time not burdening the central system unnecessarily. Thus we have two tasks at the 'user' end - a forms handling task which interacts with the operator, and a line handling task. The two tasks are identical in central and remote nodes, the Network Services Protocol layer of DECNET providing the transparency required. This transparency is one of the tremendous benefits provided by the layered design approach of network packages such as DECNET.

Graphics Work Station

No detailed functional requirements exist yet for this subsystem. Generalised functions include the following:-

+ Local development of user application programs in any of the programming languages available on the system, viz Fortran, Basic and Macro.

+ Access to the central database, from programs and interactively.

+ Comprehensive statistical manipulation subroutine library.

+ Ability to produce hardcopy graphical output by means of a plotter, with convenient development of graphics programs using a graphics CRT during the development phase.

These capabilities will be used to support modelling studies of furnace behaviour and performance, using data from the historical database, as well as the production of one-shot management reports on an ad-hoc basis.

Central System

It is difficult to decide just how much detail to present in the case of the central system. It would be easy, if perhaps rather tedious for author and reader, to present many tens of pages of descriptions of what happens in it, and while a tremendous amount of activity does take place there, much of that activity is fairly conventional and unremarkable in the context of distributed systems. It was decided therefore to confine ourselves to highlighting some of the more interesting or important aspects, without going into much detail.

The core of the central system is of course the historical database. It is a single database, with the data in it falling clearly into two categories, viz time-related and non-time-related. Time-related data includes such things as the scanner three-minute data frames, charge data from the FOX 2/30, analysis data from the spectrograph and operator entry, and operational data. Non-time-related data includes such things as process point descriptions, addresses, conversion constants, data entry format descriptors, and so on. Data from all categories is expected to take up some 80 Mbytes of storage.

Archiving of data to tape takes place automatically at 04h00 every day. At that time, the previous day's data is spooled off to tape in a format which permits simple reloading back to disk for archival data extraction. After that, data older than 40 days is deleted from the online historical database.

It was pointed out earlier that the dual Winchester disks were used in a 'cold' standby mode, so that if one fails, the system can be restarted manually using the other disk. The 'spare' Winchester is to be used in between times to support the archival data recovery function. A database structure identical to the online database will be kept on it, into which archive data can be loaded from tape, for extraction of long term historical data.

A specific requirement of the system is very rapid response to requests for operator displays. The approach taken to achieve this is to store on disk a complete image of each display format, complete with live process data. These pictures will all be updated every time a scanner frame is received. This approach is only valid as long as the number of different displays remains small. One benefit of this approach is that it provides a convenient place to put the rather complex derived variable calculations, together with the monitoring of these against various limits. Also located in this subsystem will be the preprocessing required to support the longer term statistical calculations (daily, weekly etc), and the preprocessing required to augment the query package capabilities - standard deviation calculation for example.

Scanner support includes modules to do the following

+ Receive the three-minute frames and enter the data into the historical database, and trigger the picture refresh module.

+ Receive the data packets describing alarm messages and translate and print them. Trigger the display susbsystem to update the diplayed alarm state.

+ Extract point descriptive records from the database and construct the tables required by the scan task.

+ Extract point descriptive records from the database and construct the tables required by the investigational scan task. This function must be particularly easy to drive so that relatively inexperienced personnel can use it.

+ Transmit control tables to and receive data from investigational scan task.

+ Maintain synchronisation between the time of day as held in the central system and in the scanner systems

Support for the data entry systems includes the following

+ Retrieval of forms definition records from the database

+ Retrieval of check limits and parameters from the database

+ Receiving of completed reports and storing in the database

CONCLUSION

We have attempted in this paper to give an over-view of what we consider to be the more interesting aspects of this particular computer system in the context of distributed systems. Thank you for your attention.

THE SYNERGISM OF MICROCOMPUTERS AND PLCs IN A NETWORK

I. Brown and E. F. Bosch

Persetech (Pty) Ltd., South Africa

ABSTRACT. The distinguishing features of a programmable controller (PLC) are discussed while the use of microcomputers or minicomputers together with PLCs is justified. A number of makes of PLC are compared. The comparison includes the communication features as well as the process control features of the PLCs. High availability and high reliability configurations are also discussed as are aspects of the man machine interface. An example of a system using two PLCs and one microcomputer is given.

KEYWORDS. Process control, programmable controller, programmable logic controller, distributed system.

1. INTRODUCTION

Programmable controllers are now being used in roles more sophisticated than their original role of mere sequence controllers. They are currently being supplied with standard three-term loop controllers enabling them to handle applications up to now reserved exclusively for the traditional process control products. This tendency, together with the general compexity of programmable logic controller systems requires a more complex man-machine interface. It is in this field as well as data logging and trend analysis that the micro computer can be successfully applied.

This paper will discuss what is meant by a programmable logic controller. (The acronym PLC will henceforth be used for programmable controller as well as programmable logic controller) The advantages and disadvantages of micro computers or minicomputers in this field will also be discussed. It will then compare the networking features of a few families of PLCs using a model by Gertenbach of a layered communication structure which includes both the process control as well as the communications features of the system.(ref 1) Finally, as a conclusion, an example will be given showing that in order to achieve meaningful man-machine communication on a moderately complex system one is forced to use minicomputers or microcomputers as the PLC software is at this stage not sufficiently flexible or compact.

2. WHAT IS A PLC?

The original description of a PLC as a relay circuit emulator is no longer valid as computational and data handling features have been added which are unrealisable in a relay circuit. Programming languages other than the relay ladder diagram are also being offered. The distinction between a PLC and other computers is somewhat vague. Gould Incorporated define it as "a solid state device which has been designated as a direct replacement for relays and 'hardwired' solid state electronics, whose logic can be changed without wiring changes". (ref 11)

The PLCs now offer process control features, but this is not their prime function. We should be on our guard against the over zealous "PLC Man" who implements a process control system using PLC hardware. It is an unfortunate fact that the industry will be littered with home grown solutions until the lesson is learnt that it is unnecessary to re acquire the decades of experience gathered by the traditional process control industry.

We feel that the characteristic features of a PLC are:-

a) It is a machine which handles predominantly discrete digital inputs and outputs.

b) The machine is designed to include a robust input output

system directly switching industrial voltages and currents.

c) The machine uses interpretive languages which allow code to be easily modified during commissioning and final operation of the system.

d) The hardware should function in an unconditioned environment without moving components such as fans.

3. WHY USE A MICRO OR MINI COMPUTER

Using a microcomputer in conjunction with a PLC has the following advantages:-

a) One may provide an easier and more flexible formatting of displays and print outs.

b) Mass storage on magnetic discs permits the recording of histories, trends, and the accumulation of operating statistics.

c) One may use a wide variety of high level languages to implement complex manipulations and data displays and printouts where necessary.

d) A number of software packages are available for certain industries. These may be purchased and then interfaced to the outside world through PLCs provided that the required communications layers have been developed for the equipment.

Using microcomputers with their associated peripherals in the PLC environment also has its drawbacks which can be listed as follows:-

a) The hardware is not as robust as the PLC hardware. It usually requires an air-conditioned, dust-free environment.

b) It may not have the same battery back-up as the PLCs and,

c) The hardware is not normally as modular as that of the PLCs.

4. USING A MODEL TO COMPARE A FEW PLC FAMILIES

In this paper we are concerned with PLC equipment supplied as standard hardware from the respective vendors. Any networking features discussed are therefore possible without any custom hardware or software interfaces. The model used as a basis for comparison of the PLCs will be that developed by Gertenbach (ref 1) for distributed computer process control systems.

The above model is based on the different layered architectures used to describe inter computer communication in public data networks. The levels of description of this model are described in figure 1. This figure represents one of many pillars, each one of which stands on a processor as it is configured in a horizontal plane to constitute the network topology. In a process control distributed system the plant interactions behave as if they were a communications network in parallel with the communications designed into the system. It is therefore necessary to include a communications leg as well as process leg in each pillar which stands on a processor which also interfaces directly to the plant.

Apart from a few exceptions, the PLCs offer services in only the lower few levels. It is the user's responsibility to provide the other layers in his application program. One of the exceptions is the Modicon Modway public industrial local-area network. It is on a higher level than Modicon's Modbus system which is one of the general communications facilities provided. Modway is, strictly speaking, not exclusively a PLC product even though it is manufactured by a PLC vendor. It has not as yet been released as it is still undergoing site testing.

A number of small PLCs do not offer any form of communication. These PLCs are aimed at the low end of the market and are very cost effective in their application. These should not be excluded as their cost effectiveness as small sequencers etc. justifies their existance in a simple network with a 6 wire interconnection setting and sensing relays for start, stop, acknowledge, and complete signals. We will refer to these as the Basic range. They include the Tempatron, Omron, and the Kostac SR-40 ranges.

Other PLCs have serial links but these cannot be regarded as communications ports because of limited software support. These links will, however, handle limited primitive communication. We refer to these as the Basic communications range. They include the Festo and the Sprecher and Shuh ranges.

We will now compare the PLCs level by level.

4.1 LEVEL -1 PHYSICALLY CONTROLLED MEDIUM

4.1.1 LEVEL -1 COMMUNICATIONS LEG - THE MEDIUM

The basic PLCs may use a 6 wire communication line for a simple start-stop protocol. This is obviously wired point to point.

The "basic PLCs with communications" as well as the Automate, the Siemen's Simatic, and the TPC 8000 PLCs use 4 wire point to point communications. The GEM 80, Modbus, Struthers Dunn, and Allen Bradley systems use 4 wire multidrop systems. The GEM 80 and Struthers Dunn Systems provide off-the-shelf fibre optic couplings. The Modway system uses a coaxial cable multidrop system.

4.1.2 LEVEL -1 PROCESS LEG THE PLANT

This level describes the plant being controlled by the PLC systems. It is not considered in this comparison.

4.2 LEVEL 0 SIGNAL LEVEL

4.2.1 LEVEL 0 COMMUNICATION LEG DATA CIRCUIT EQUIPMENT

This level refers to features or equipment which adapt the higher level signals to the communications medium and vice versa. Equipment such as modems and line drivers and receivers falls in this level which is summarised in table 1.

4.2.2 LEVEL 0 PROCESS LEG TRANSDUCERS & ACTUATORS

This level describes the transducers and actuators used by the systems to control the respective plants. These do not form part of PLC systems as they are currently marketed, with the result that this level is not included in the comparison.

4.3 LEVEL 1 - BIT LEVEL

4.3.1 LEVEL 1 COMMUNICATION LEG - THE DTE INTERFACE

This is comprised of the physical, electrical, functional, and procedural facilities required to establish, maintain, and disconnect the physical link between the various processors or data terminal equipment.

The functions of the PLCs in this layer are compared in table 2.

4.3.2 LEVEL 1 PROCESS LEG THE PROCESS INTERFACE

This level together with the system level is a major area of concern when selecting a PLC for a particular application with PLCs. This level concerns itself with the conditioning of the input and output signals. In PLC systems this concerns the input voltage sensors and the output contacts, their physical connection facilities and safety features. This level is especially important for the smaller devices as they are usually used in simple applications where the programs are almost trivial thus making the programming language less important. The features of the PLCs in this layer are compared in table 3.

4.4 LEVEL 2 THE FRAME LEVEL

4.4.1 LEVEL 2 COMMUNICATION LEG - DATA-LINK CONTROL

This level performs the setting up of links, building the frames, positive and negative acknowledge, flow control and error control. The frame level function must be performed by the user application program in most cases. The Modway system and a possible development by Siemens on the proposed IEEE process control specification are exceptions to this.

The limited services provided at this level by a few PLCs are given below.

4.4.1.1 GEM 80

The system is configured by setting up data in the preset data tables. A PLC can be configured as a master or a slave. It can communicate with other PLCs or with a computer. The computer may be either the master or the slave. The master PLC transmits the contents of its 32 word output buffer into the slave's input buffer. The slave then responds by transmitting the contents of its output buffer into the master's input buffer. This can be free running synchronised with the PLC scans, or else it can be dependent on the setting of a flag by the user.

4.4.1.2 Modbus

A vendor supplied package in the form of ladder diagram code allows the user to access event and state information in other PLCs. A FORTRAN package is available to allow any computer to communicate with the PLCs. The computer is the master in this case and uses a PAR protocol.

4.4.1.3 Simatic

As with Modbus a software function allows two PLCs to communicate with each other. This is however only point to point communication. Any of the two PLCs may initiate a message while a priority circuit handles clashes. A software package is also available for computers communicating with these PLCs.

4.4.1.4 Struthers Dunn

A computer may communicate with the PLCs. The PLCs are however always the slaves. A PAR protocol is used in the central polling computer. Error states are, however, passed back as a negative acknowledge in certain circumstances. Communication between PLCs is possible but is limited to the passing of 32 bits in either direction by a "system interconnect track". This is a serial link which performs the equivalent of connecting 32 pairs of wires between output terminals on the one unit and input terminals on the other and vice versa.

4.4.2 LEVEL 2 PROCESS LEG
INPUT/OUTPUT CONTROL

This level describes the control hardware and software which controls the input output system interface bus. This is implemented in all PLCs but is transparent to the user in all systems.

4.5 LEVEL 3
THE PACKET LEVEL

This level is responsible for the reliable transmission of data and control information between processors and from the processors to their I/O elements.

4.5.1 COMMUNICATION LEG

The packet transmission level provides a logical connection to transmit packets between processors situated in different processors. This is done by means of virtual circuits or datagrams. A very simple datagram is used in most cases.

4.5.2 PROCESS LEG
CENTRALISED I/O CONTROL

This level ensures that the correct output contacts are closed or that the correct inputs are read. In a PLC it copies the contents of the output memory locations into the interface blocks, and the status of the interface blocks into the memory locations. This is transparent to the user in all systems.

4.6 INPUT OUTPUT BUS OPTIONS

The PLCs offer a variety of input output bus configurations. The remote I/O serial busses should also be mentioned in the comparisons of the physically controlled medium, signal level, and bit level. We describe the serial busses separately, however, so as to simplify the comparison.

This section describes the interconnection between the PLC I/O modules and the central processor modules. The interconnections may be divided into 3 levels:-

a) The mounting rail bus.
(This links the modules in an I/O chassis)
b) The inter-rail bus
(This links processors and closely situated chasses)
c) The remote I/O connection
(This links processors and chasses separated by large distances.)

The mounting rail bus usually consists of a flat cable with the usual flat cable connectors at the various slots. It is also sometimes a printed circuit card.

The inter-rail bus is usually a flat cable with a maximum end to end distance of about 10 metres. This arrangement is in our opinion, a weak point in most systems. The Struthers Dunn System which uses the IEEE 488 bus, and the Allen Bradley system which regards all I/O as "remote", are exceptions to this.

The remote I/O or serial I/O as it is also known is a feature which can lead to enormous savings in cabling costs. In many cases geograhically separate I/O modules still belong in one functional group and do not justify distributed processing. They can now be centrally controlled but the various groups of signals are interconnected by a single coaxial or twisted pair cable instead of the myriad of cables used in the conventional approach. This can, however, be a disadvantage as designers could be tempted to centralize the control of plants ideally suited to distributed computer control. The remote I/O features of a few PLC families are compared in table 4.

4.7 HIGH AVAILABILITY AND HIGH RELIABILITY

The redundancy configurations supplied by different vendors involve features at a number of levels of description in the Gertenbach model. These features are compared separately, again for the sake of simplicity.

Four systems are described, three hot standby, and one safety circuit. The most sophisticated hot standby configuration is that offered by Modicon, followed by Allen Bradley, while the GEM 80 has a simpler bus switch. Siemen's Simatic system offers a dual-processor high-reliability failure detection system on one of its smaller PLCs which has been approved for safety applications by the Bavarian Standards Organisation. The three hot standby systems are described in figure 2.

The Modicon redundancy is achieved by the redundancy supervisor which controls a remote I/O bus switchover unit. This switchover unit connects the remote I/O bus to either of two identical processors.(Only the largest model in the family) It also switches the serial links going to external computers, terminals, and other PLC's into the respective processor. The PLC which is currently controlling the I/O bus and communications busses in known as the active unit. The other PLC is known as the standby unit. The system is symmetrical, namely, the outside world does not know which processor is controlling it. A redundancy supervisor performs regular checks on both processors, it also checks that the programmes in both processors are identical. It synchronises the scans of the two processors and passes the entire database of the active unit to the passive unit at the end of each scan. The redundancy supervisor instructs the switchover unit to change processors as soon as it detects a fault in the active processor. If a fault is detected in the passive unit as well, then the redundancy supervisor ensures that all outputs revert to their safe state. A possible disadvantage of this is that one would be obliged to connect a large number of I/O points onto the remote I/O bus in order to justify the expense of the two large processors, the redundancy supervisor, and the switchover unit. The I/O network failure rate obviously increases with this increased complexity.

The Allen Bradley PLC 3s have a feature which allows two processors to be connected to the I/O bus simultaneously. A processor is designated active or standby. If a processor is in the standby mode it can sense inputs but it cannot stimulate outputs. A data link is also necessary between the processors in order to continously transfer the status of counters and other integrating devices from the active to the standby processor so as to provide a smooth switch over.

A proposed GEM 80 bus switch, switches the local data highway to either of two processors depending upon the status of the PLC watchdog circuits. This does not have the same built-in features to facilitate bumpless switchover as in the Modicon.

The Siemen's programmable controller for safety oriented controls handles a maximum of 16 safety actuators and 32 safety transducers coupled to corresponding input modules in each of two systems. The two units clock synchronously while executing identical programs in parallel. Two comparator units check the binary signals on both bus systems for equality after each statement. All safety circuit outputs are immediately switched off if a difference occurs between the signals on the two busses.

4.8 IMPLEMENTATION ISSUES

The PLCs generally provide the communications primitives through the routines described in level 2. The primitives consist in this case of the setting of certain data words equal to the target processor physical address, the address of the data in the target processor which is to be read or changed, the command such as read or write, the number of words to be transferred, and the address of the source data inside the local PLC. In the GEM 80 for example the data transmission primitive consists of a rung which moves the required block of data into the transmission buffer.

A essential feature is the provision of all the usual operating systems facilities. In the PLCs this takes the form of the instruction interpreters. If one bears in mind that even in PLC systems the cost of the application software exceeds the hardware costs one appreciates that this layer must be weighted heavily in any selection process where the program is larger than about 8 kilobytes.

As the early PLCs were aimed at the direct replacement of relays their interpreters were designed to behave exactly as if they were relay circuits. This required the hiding of the serial nature of the operation from the users. Some machines still prohibit the user from outputting to a bit or "coil" more than once. Many machines also do not allow the user to employ the fact that the coils are merely 1 bit in a 16 bit word. "Coil" data and "register" data is not easily interchangeable. The early ladder diagram languages are not suited to the larger applications employing PLCs with 32k bytes of user memory and more.

The Siemen's Simatic system employs subroutines and function modules while the Allen Bradley PLC/3 allows subroutine calls. We hope that the other manufacturers will follow this example and bring PLC programming back into the computer age. The Simatic system also allows the user to use three different languages when building a system. These are the control system flowchart, the ladder diagram, and the statement list. The statement list is a hard to document assembler like language which is, however, very flexible and ideally suited to the programming of the function modules. These function modules will usually be regarded as black boxes by the end user who need only work in the higher level languages when making the inevitable modifications to his system. The two high level languages do not have many differing features. They were merely designed for users coming from the computer industry on the one hand and from the relay control industry on the other.

The GEM 80 family has expanded their ladder logic to inclule linear functions in the ladder diagram. In our opinion this makes the most readable ladder logic on the market. While the GEM 80 software has not progressed as far as subroutines and data blocks they have provided very useful block statements where blocks of code can be enabled or disabled as if they were subroutines. These blocks have a "start of block" and "end of block" statement as opposed to the jump statements used in most other PLCs.

A number of manufacturers are now providing parallel processing in the form of smaller processors with limited I/O which can be plugged into the main chassis. Although these units are independent processors with their own CPU and memory, they generally use the same power supply as the main processor and are connected to the main processor by means of a closely coupled bus. We feel that at this stage these units should be regarded as a means of increasing processing speed rather than as a distributed processor.

5. MAN-MACHINE INTERFACE

Gertenbach lists the four main man-machine interface functions in a real time system, namely:

a) The generation of process logs and alarm reports.

b) The selection of process parameters for control or display.

c) The display of process parameters in digital, analog, or graphic format.

d) The actual modification or setting of process parameters.

It is at this level that the power and flexibility of microcomputers can transform the PLC system into a friendly system which the operating staff will use with ease.

A number of PLCs offer graphic colour displays for very effective mimic displays. While these mimic displays are a good indication of plant status they allow only limited operator interaction. The setting of setpoints, time constants, maximum an minimum values etc. can be very cumbersome using these machines. The GEM 80 allows one to use BASIC in a parallel processor which plugs into the main chassis. Reasonably complex operator communications programs are generally extremely large at the best of times, however, when they are to be written in BASIC or ladder logic they become quite unmanageable. In this situation one should seriously consider the established compiler languages such as PL1, PASCAL or FORTRAN. The microcomputer functions will not replace the PLC features but will in general provide valuable enhancements.

6. AN EXAMPLE OF A MICROCOMPUTER PLC COMBINATION.

A system installed by our company involved the control of a batching plant feeding a variable mix of eight materials into 3 ferro silicon furnaces. A multipath transport system consisting of 36 conveyors feeds the material into a total of 21 charging bins on top of these furnaces. The batches are corrected for overshoot and undershoot. A feedback control system was implemented to ensure perfect layering of the material. The system was implemented using two PLCs. One to control the batching process, and another to control the transport system. These PLCs in turn communicated with a microcomputer with a flexible disc, a printer, and a video display terminal.

The system, although it was fairly complex in relation to other batch weighing systems, was small when compared with some large multiple PLC applications. It was found however, that without a microcomputer it would have required much more operator skill and patience to perform the required data entry and modification. The size of the logging task alone would have justified the addition of the microcomputer.

The operator changed the parameters using a dialogue in which the computer presented menus and prompted him for response. All data entered was checked for reasonableness. The functions of the operator were to change material recipes for each of the furnaces, the percentage moisture for each day bin, and the quality figure for each carbonaceous material. Furnaces had to be switched in or out, the operating delays needed to be changed to cope with problems such as belt slippage, while alarm timeouts also required adjustment occasionally. Besides this, procedures were implemented to provide semi-automatic calibration of the weigh bins and of the discharge feeder rates. The logging task printed masses thrown to each furnace every hour, shift, and day besides providing a detailed log of all operator activities in order to be able to analyse furnace disturbances should they occur.

Without the computer it would have been impossible to implement these functions which we feel were the main reason for the high level of operator acceptance and the fact that with a few hours training one could leave an operator new to the computer system in charge of the batching system feeding three large furnaces with a total power consumption in the order of 64 megawatt.

7. CONCLUSION

Multiple PLC systems using standard hardware and software are now easily implementable. With one or two exceptions the PLC manufacturers have, however, not developed a standard interfacing strategy where one can easily mix makes of PLC and PLCs and computer's.

Notwithstanding the rapid advance in PLC hardware and software it will still be necessary to use mini and micro computers in order to perform easily understood and configurable man machine communication.

8. REFERENCES

1 Gertenbach W P
"AN INVESTIGATION INTO THE ORGANISATION AND STRUCTURED DESIGN OF MULTI-COMPUTER PROCESS-CONTROL SYSTEMS"
Phd Thesis Dept of Electronic Engineering
University of Natal

2 Gertenbach W P
"AN HIERARCHICAL MODEL FOR DISTRIBUTED MULTICOMPUTER PROCESS CONTROL SYSTEMS"
5th IFAC Workshop on Distributed Computer Control Systems
May 1983

3 Allen Bradley Product Data Bulletin 1775-900
"PLC-3 PROGRAMMABLE CONTROLLER"

4 Allen Bradley Product Data Bulletin 1775-903
"1771 I/O RACK USED WITH PLC-3 CONTROLLER"

5 Allen Bradley Product Data Bulletin 1775-909
"PLC-3 COMMUNICATION ADAPTOR MODULE"

6 Allen Bradley Publication 1775-803
"PLC-3 PROGRAMMABLE CONTROLLER BACKUP CONCEPT MANUAL"

7 Control Systems Division of Reliance Electric Publication AM005b
"THE AUTOMATE 35 PROGRAMMABLE CONTROLLER"

8 Festo Unnumbered Publication
"WHY THE FESTO FPC REPRESENTS THE MOST COST EFFECTIVE SOLUTION WHEN CHOOSING A PROGRAMMABLE CONTROLLER"

9 GEC Electrical Projects Preliminary Data 8927/80
"GEM 80 8927 FIBRE OPTIC TRANSCEIVER"

10 GEC Electrical Project Preliminary Data DP34
"BASIC I/O CHANGE OVER UNIT DP4"

11 Gould Inc. Modicon Division publication 40052
"SUMMARY DESCRIPTION MODICON 484 PROGRAMMABLE CONTROLLER "

12 Gould Modicon Division publication 40053
"SUMMARY DESCRIPTION MODICON 584 PROGRAMMABLE CONTROLLER"

13 Gould Inc. Modicon Division publication ML-J211-USE
"J211 REDUNDANCY SUPERVISOR USER'S MANUAL"

14 Gould Modicon Division publication 40086
"CONCEPTUAL DESCRIPTION MODWAY PUBLIC INDUSTRIAL LOCAL AREA NETWORK"

15 Gould Modicon Division publication 40131
"INDUSTRIAL COMMUNICATIONS SYSTEM MODBUS"

16 Gould Modicon Division publication 40151
"MODBUS SYSTEM PLANNING USERS MANUAL"

17 Koyo Electronics Industries Co. Ltd Publication JAN 82 0720-0-5CH
"KOYO PROGRAMMABLE CONTROLLER"

18 Omron Tateisi Electronics Co. Publication 882-1X
"SYSMAC PO SEQUENCE CONTROLLER"

19 Omron Tateisi Electronics Co. Publication 382-15M
"SYSMAC POR PROGRAMMABLE CONTROLLER"

20 Omron Tateisi Electronics Co. Publication 682-3M
"SYSMAC-MIR PROGRAMMABLE CONTROLLER"

21 Siemens Product Bulletin E324-P-82/10
"SIMATIC S5-110F PROGRAMMABLE CONTROLLER FOR SAFETY CIRCUITS"

22 Siemens Product Bulletin E324 /1879-101
"SIMATIC S5 670 PROGRAMMING UNIT"

23 Claus Becker, Gerhard Bier, Heinz Bischoff, Gerd Hinsken
"S5-150S PROGRAMMABLE CONTROLLER FOR NEW APPLICATIONS IN THE UPPER PERFORMANCE RANGE OF THE SIMATIC S5 AUTOMATION SYSTEM"

24 Siemens Catalog ST 51 1982
"S5-110 PROGRAMMABLE CONTROLLERS"

25 Siemens Catalog ST 55 1981
"S5-150 PROGRAMMABLE CONTROLLERS"

26 Sprecher and Schuh Publication 86 23
"PROGRAMMABLE CONTROLLER SESTEP 530"

27 Struthers-Dunn Systems Division Part No 79669
"DIRECTOR 4001 PROGRAMMABLE CONTROLLER SYSTEMS MANUAL"

28 Tempatron unnumbered publication
"TPC 8000 SERIES OF PROGRAMMABLE CONTROLLERS PRELIMINARY SPECIFICATION"

	LIMITED DISTANCE MODEM	FULL MODEMS	RS232 OR 20MA	RS422
AUTOMATE		X	X	X
ALLEN BRADLEY	X		X	
GEM 80	X		X	
MODBUS		X	X	
MODWAY		X	X	
SIEMENS	X		X	
STRUTHERS DUNN	X		X	X
TEMPATRON			X	X

TABLE 1 SIGNAL LEVEL

	MAXIMUM NUMBER OF STATIONS	POLLING SINGLE MASTER	MESSAGE INITIATED FROM BOTH SIDES	MULTIPLE MASTER "TOKEN"
AUTOMATE	2			
ALLEN BRADLEY	64			X
GEM 80	7			
MODBUS	247	X		
MODWAY	250			X
SIEMENS	2		X	
STRUTHERS DUNN	31	X		
TEMPATRON	2	X		

TABLE 2 BIT LEVEL

PLC FAMILY	AC INPUTS (110v, 220v)	AC OUTPUTS (110v, 220v)	DC INPUTS (10v - 50v)	DC OUTPUTS (10v - 50v)	OUPUT STATUS INDICATION	FUSES ON OUTPUT CONTACTS	REPLACE I/O MODULES WITH POWER ON	REASONABLY SIZED SCREW TERMINALS	REMOTE INPUT/OUTPUT	REPLACE OUTPUT FUSES WITHOUT REMOVING MODULE	REMOVE I/O MODULE WITH-OUT DISTURBING WIRING	INTRINSICALLY SAFE	MAXIMUM NUMBER OF DIGITAL INPUTS	MAXIMUM NUMBER OF DIGITAL OUTPUTS	MAXIMUM NUMBER OF ANALOGUE INPUTS	MAXIMUM NUMBER OF ANALOGUE OUTPUTS
AUTOMATE	X	X	X	X	X	X		X		X	X		2048*	2048*	64*	64*
ALLEN BRADLEY	X	X	X	X	X	X		X	X		X		4096	4096	512	512
FESTO	X	X	X	X	X								128	128	16	0
GEM 80	X	X	X	X	X	X		X	X		X		512*	512*	192*	192*
KOYO SR 40		X	X	X				X	X		X		24	20	0	0
MODICON	X	X	X	X	X	X	X	X	X		X	X	8192*	8192*	1000	1000
OMRON	X	X	X	X	X			X					512	512	0	0
SIEMENS	X	X	X	X	X			X	X	X	X	X	18944	18944	192	192
STRUTHERS DUNN	X	X	X	X	X	X	X	X	X				256*	256*	32*	32*
SPRECHER & SCHUH		X		X	X								512	512	32	32
TEMPATRON	X	X	X	X	X								56*	56*	0	0

* SUM OF INPUTS AND OUTPUTS IS GIVEN

TABLE 3 COMPARISON OF PLC I/O FUNCTIONS

	MEDIUM				MAX DATA RATE K band	MAX DIST-ANCE METRES	MAX I/O AT END OF EACH LINE	MAX NUMBER OF ENDS PER BUS	FAIL SAFE ON LINK FAILURE	REDUN-DANT LINK
	CO-AXIAL	BI-AXIAL	4 WIRE	FIBRE OPTIC						
ALLEN BRADLEY		X			57	3000	128	16	X	
AUTOMATE			X		256	300	128	1		
GEM 80			X	X	9,6	300	1024	7		
MODICON	X				1200	9000	512	16	X	X
SIEMENS			X		9,6	1000	1024	1	X	
STRUTHERS DUNN			X	X	56	1200	32	1	X	

TABLE 4 PLC REMOTE INPUT OUTPUT FEATURES

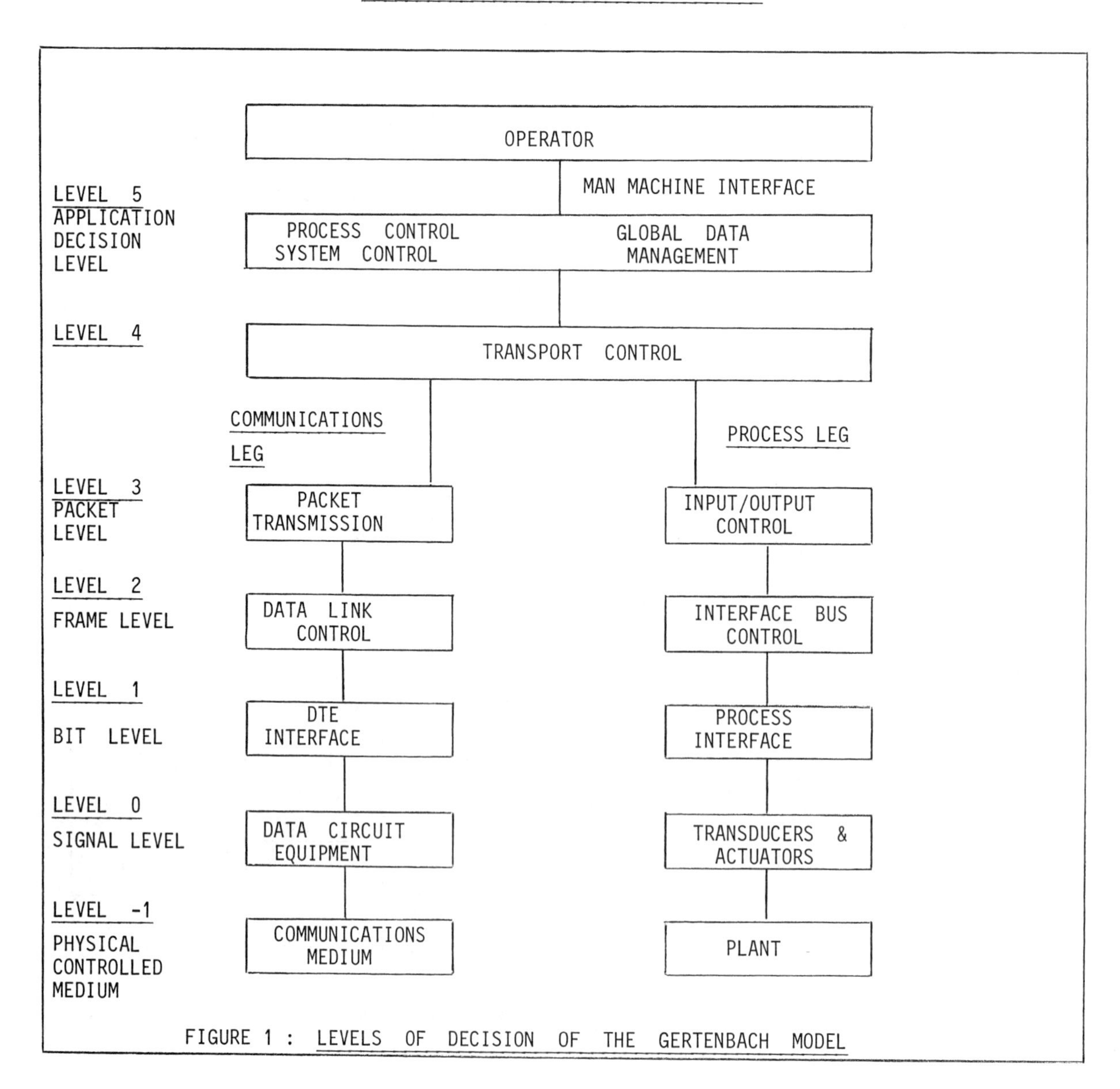

FIGURE 1 : LEVELS OF DECISION OF THE GERTENBACH MODEL

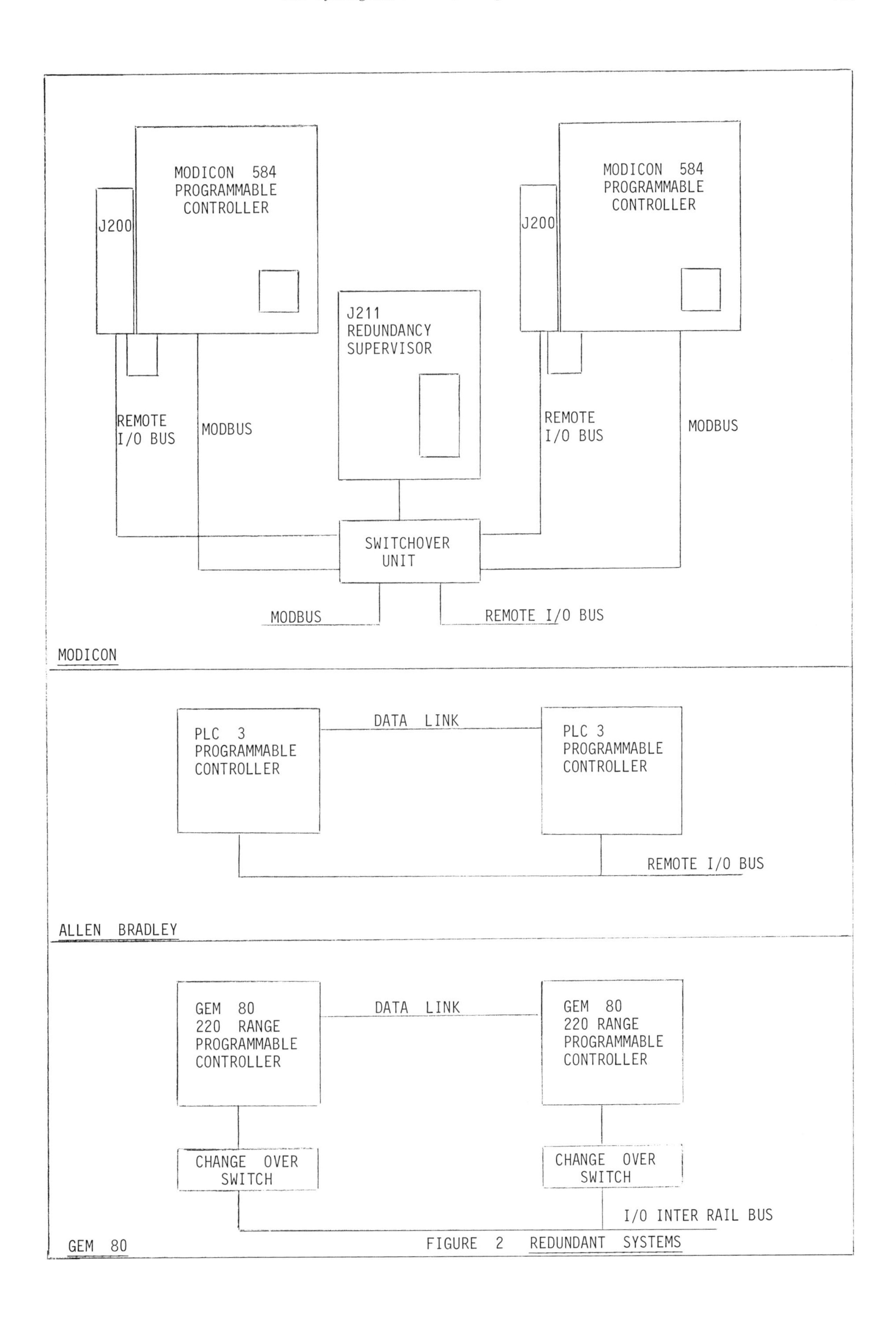

FIGURE 2 REDUNDANT SYSTEMS

DISCUSSION

Peberdy: This has been a very relevant and interesting session which has followed naturally from earlier discussions. We started the workshop with fairly esoteric papers - we had one earlier application paper by Mr Imamichi which was very interesting but the system architecture, and the hardware and software that were used in that application employed a lot of special effort. This application session is interesting and relevant because the papers solve today's problems using today's off-the-shelf technology. So maybe we have come down to earth, hopefully with not too much of a bump. Also I think that the applications which have been discussed are perhaps typical in terms of the number of applications of this size and scope, certainly in this country and probably elsewhere as well.

LaLive d'Epinay: I would like to ask the speakers as to what the estimate of engineering costs would be if they were to add a second identical man-process interface, such as a graphical station, to the systems which have been described.

MacDonald: To add a data entry station to the system would merely be a question of connecting the appropriate hardware. The hardware which we are using allows up to 250 nodes on the multi-drop cable. Provided we assume that the bandwidth of the data link can support the number involved, there are no engineering costs involved other than just physically plugging the station in.

Gertenbach: Mr Byrne said that most of the systems that we can consider as being distributed are not strictly distributed but should rather be referred to as resource sharing with decentralized functions.

Byrne: The point I was trying to make was that it is my belief that many of the decisions about the level at which distribution is to be made are trivial. If I could use an example of that, you can take a process and divide it into 100 units in terms of node computers. You don't have the choice of dividing it into either 50 or 200s. Would it be better to divide it into 200 rather than 100? I am not really talking about geographical distribution, I think that is totally irrelevant as well. I think you will find that in general terms the majority of proprietary distributed control systems that are installed are in fact not geographically distributed. You will find that in virtually 90% of the cases they are in fact installed in a control room. So I think geographical distribution is really irrelevant.

Rubin: In many ways I share the philosophy which Mr Byrne has expressed. I feel that distributed control is essentially an operation of delegation in the same way as a manager must delegate to subordinates and direct his attention to strategic planning instead of operational matters. In the same way when we talk about a distributed control system we are talking about a manager who is doing high level management, being located in one position, and distributed boxes within the plant doing the operational management. I agree that this delegation gives you, in fact, tolerance to local interruption of communications or whatever else may occur. Once a job has been delegated it can be performed without minute-to-minute supervision.

Byrne: Often the various levels of distribution seem to be totally arbitrary, but of course they are not. The only common link that they have is the inflexibility of the vendor supplied equipment. No amount of effort will permit you to squeeze a nine-loop system into an eight-loop controller. You know when you design the system that there is always that constraint and that is why the level of distribution is usually totally arbitrary to the plant.

Kopetz: I would like to get the terminology straightened out, because I think if you talk about distributed systems you could think of distribution in 3 domains. Distribution in the hardware domain (which can be locally or geographically distributed), distribution in the control domain and distribution in the data domain. Now there are some people who call a distributed system only distributed when it involves all three domains and others who call it distributed if any one of the domains is distributed. I would like to propose a fourth definition. As far as I see it a distributed system must necessarily be redundant because if there is not redundancy then it is simply a resource sharing system.

I would like also to raise the question relating to the system on the blast furnace. When you look back at the effort you spent on implementing the system, how much time did you spend on the application software, and how much on the communication software?

MacDonald: Because we are using a standard package there has been virtually zero effort involved in the communication software. A major objective of the whole project was to use, as far as possible, standard off-the-shelf components for both hardware and software, so for that reason we have made use of a standard off-the-shelf communication package which provides a very high level of task-to-task, task-to-remote file, remote-terminal or remote-computer, communications.

Kopetz: This then is DECNET? In my experience it takes a large amount of time to get involved with DECNET, I certainly would not consider it to be trivial.

MacDonald: It certainly is a non-trivial effort, but it is small by comparison with the effort which goes into application software generation. I would also like to refer back to comments made about my presentation. I did not say that you should not control loops with PLCs. What I was saying is that one should be on guard against the in-house developed systems that use large PLCs to perform analog control functions which could be better performed by a proprietary system. In fact PLCs do have a certain limitation in analog control in that they, in performing say a PID loop, do not use absolute time. In practice they use the scan time - which can vary. In most cases one would get away with it but there are possible situations where your program may have to jump through a different loop at a particular stage and your control would be faulty. Most PLCs have a real-time capability, no matter what form it comes in. They can, in fact, synchronize things with maybe a jitter on the timing, but you could synchronize to perhaps 1, 5 or 10 seconds independently of what the cycle time is. Finally in my definition of a distributed system I feel that if you have got more than one computer, no matter what the motivation for it was, then you have a distributed system.

Maxwell: I would like to refer to the previous speaker's papers. I would like to know what the justification for their various data collection systems was and also what is the future for control in their applications?

Stevens: We did not motivate the system in practice - the system was motivated by our production department. Based on their examination they felt that they could significantly improve the fuel rate on the furnace and, as the main fuel input is coke, if you can save on the charge, (approximately 50% of the burden) you can save significant quantities of coke. The feeling amongst our production engineers was that with better monitoring of the furnace and a better understanding of things through a more scientific examination of the performance, then it would be possible to bring down our coke rate from about 520 kg per liquid ton of iron, as it is today, to about 480. This saving would pay for the system in less than a year. So this was really the motivation for the collection - it was essentially a research tool and as such there is probably an element of gamble associated with it. We certainly do see the possibility of closing the control loop in the future, either by extension of the system or by the provision of information into the existing control computer.

Brown: Our experience has been that there has been a saving, achieved after installation, which has proved that the installation was worth the money spent. I am not certain whether the customer had done a full motivation beforehand, but by more even loading of the furnace they achieved availability far higher than they ever experienced in their entire history. This meant that they had fewer breakages of electrodes, which are extremely costly and can lead to a furnace being put out of action for at least a day.

AUTHOR INDEX